Tower
London
Southwark
Millennium
Blackfriars
Waterloo
Hungerford
Westminster
Lambeth
Vauxhall
Chelsea
Albert
Battersea
Wandsworth
Putney
Hammersmith

CROSS RIVER TRAFFIC

A History of London's Bridges

Chris Roberts

Granta Books
London

Granta Publications, 2/3 Hanover Yard, Noel Road, London N1 8BE

First published in Great Britain by Granta Books 2005

A CIP catalogue record for this book
is available from the British Library.

1 3 5 7 9 10 8 6 4 2

ISBN-13: 978-1-86207-800-0
ISBN-10: 1-86207-800-9

Typeset by M Rules

Printed and bound in Italy
by Legoprint

CONTENTS

On 6 March 2004 a place of love and beauty was lost to London. We shall not see its like again and this book is dedicated to anyone who ever drank, fell in love, fell over or fell out at the incomparable Dive Bar/Kings Arms, late of Gerrard Street W1, whose passing is a great loss to the community.

PREFACE

The bridges of central London are an interesting collection of Victorian and modern, sturdy and ornate, spectacular and homely. Seventeen of them allow pedestrian access: the majority are shared road and foot bridges, two are for pedestrians only, and the other is strung out alongside one of the six railway crossings. Millions of Londoners go months without seeing any of them because they travel underneath the river or never cross the Thames at all, and of the hundreds of thousands of individuals who do use the bridges every day, very few take time to wonder at the wonders they are crossing. Each bridge provides its own different and magnificent view of London, with the result that, with the obvious exception of Tower, not many people look at the crossings themselves. One of the aims of this book is to refocus attention on the bridges, many of which are beautiful architecturally and important socially. Another objective is to link the stories of London's bridges to the city and its inhabitants by showing how the bridges have helped to shape London.

'London' is a fluid concept and a decision had to be made about which crossings to include in this book. The area of the original walled city would have limited the book to only those bridges between Blackfriars and Tower; alternatively, there are twenty-seven crossings on the salty tidal reach of the Thames between Teddington and the QE2 bridge at Dartford. The Greater London Authority area goes a step further west: as far as Kingston. It could even be argued that London's modern boundary is the M25, in which case all forty-one crossings out to Egham would be candidates. This would overtop the appetites of the twelfth-century City fathers who claimed jurisdiction of the river to Staines, while others involved in the government of London had even greater ambition. The presence of the county crests of Essex and Surrey, along with their county towns of Guildford and Colchester, on Hammersmith Bridge suggests a metropolis insatiable in its passion for expansion and not sure where to stop.

Cross River Traffic – a title chosen as a nod to both Jimi Hendrix and the 1926 Royal Commission on London's Bridges – is more modest in its reach and stops at Hammersmith because that is the boundary of the old London County Council, which was the first local government body for London that covered more than the old walled city. The choice is not a deliberate snub to the good places of Chiswick, Kew, Twickenham and Richmond, all of which have fine, and in some cases historic crossings, but they did not form part of the original Victorian metropolis. The bridges that are included are presented in chronological order of construction as this best illuminates their impact on the development of London.

Like London, the book grew erratically, in stages. Bits were knocked down, chapters embanked, history bridged and tales floated together. The Thames meanders in and out of the narrative alongside other themes, both contemporary and historical, that explore aspects of London life and the city's relationship with the river. Some material relating to the Thames and the bridge builders has been placed in the appendices so that, rather than flowing off down other channels or getting undermined by tunnels, the central story is stitched together by the bridges.

Chris Roberts
April 2005

THE BURNING THAMES I HAVE TO CROSS

Old Father Thames just keeps rolling along: 5200 million litres of water, collected from 6000 kilometres of tributaries, passes over Teddington Weir every day. The river rises in the Cotswolds and flows for 344 kilometres until it reaches the ocean. Thirty million years ago the Thames flowed along a more northerly route before it joined the Rhine in the middle of what is now the North Sea. More recently, during a Pleistocene glacial period, ice sheets broke through Goring Gap and when the ice melted the Thames followed suit to take something close to its current course. The broader but shallower Thames meant, for example, that Thorney Island, the site of Westminster Abbey, was literally an island in the stream and that large sections of 'inner London' were little more than marshland, liable to flood after heavy rains or a high tide.

Marlow, the narrator of Joseph Conrad's *Heart of Darkness*, muses that the Thames valley has been 'one of the dark places of the earth'. Certainly the origin of the river's name is somewhat murky. Etymologists can only make educated guesses as to its root. Julius Caesar refers to it as Tamesis in 51 BC, but the name is much older. The Celtic root 'Tam' means dark and the pre-Celtic 'Ta' translates as turbulent flow: either way, a dark rapidly flowing river.

The river is still a haunted, uncivilized force despite the best efforts to control it, and the embankments at Tower Bridge are now fifteen metres higher than they were 2000 years ago to protect the city. These defences, along with Teddington Lock and the Thames Barrier at Woolwich, all try to regulate the surging waters. For centuries the narrow arches of the Old London Bridge performed a similar role by constricting the flow of the river and blocking the tide's advance. The Thames in London changes in depth according to the status of the tide by between five and seven metres, and the chartered depth (that at the lowest of tides) varies from nearly six metres at Tower Bridge to 1.4 metres at Hammersmith. Sewage no longer routinely flows directly into the Thames and smog is a thing of the past; the faster-flowing, embanked river of today provides a cleaning function – on hot, still summer days it helps to ventilate London.

Over the centuries the river has influenced a great many artists: Turner's Thames pictures inspired Whistler and Monet to expand the boundaries of landscape painting. It has also attracted writers, from the whimsical (Jerome K. Jerome's *Three Men in a Boat*) to the bleak, with the estuary being a metaphor for death in Charles Dickens's *Dombey and Son*. Poets and songwriters from Wordsworth to Ray

Davies have celebrated crossing it, though perhaps the most enigmatic verse is from an anonymously written folksong, the 'Grey Cock', in which the Thames appears to double as the Styx, a river of death.

I must be going, no longer staying,
The burning Thames I have to cross.
Oh, I must be guided without a stumble
Into the arms of my dear lass.

THE BURNING THAMES I HAVE TO CROSS

The Thames has served as a political as well as a physical barrier, dividing the Celtic tribes of Catuvellaunis and Atrebates and, after the Romans, the kingdoms of Wessex (Saxon) and Mercia (Dane). Even in the post-Conquest years, south of the river was seen as a potentially uncivilized, though non-specific, threat. Although there is evidence of a bridge at Vauxhall dating back 3000 years, for most people until relatively recently crossing the Thames involved either getting into a boat or getting wet. There is strong archaeological evidence of fording places at Brentford, Wandsworth and Battersea where the river could be traversed, and piers for boat crossings at Putney, Lambeth and throughout the City.

Caesar's campaigns in 54 BC and those of Claudius ninety years later used a combination of boats and fords to cross the river, but the Romans later built a bridge by their city of Londinium. When the Roman crossing fell into disuse it was 500 years before the first Saxon bridge was erected. After this bridge and a replacement were destroyed, William II, in 1097, raised a special tax to help repair the then wooden London Bridge. The monies raised were administered by a charity called the Bridge House Estates which went on to finance the first stone bridge over the Thames in

1176. Peter de Colechurch, a priest and head of the Fraternity of the Brethren of London Bridge, completed this in 1209. The Bridge House Estates continued to look after it over the next 500 years, during which time it was the only crossing on London's Thames.

London Bridge ensured that trade flowed into London, and the City elders were vociferous in protecting their hard-won rights and freedoms. The Corporation of London once controlled the river all the way to Staines, thirty kilometres away from the City, where a stone is inscribed with the legend 'God preserve the City of London', after a cash-strapped Richard I sold the Thames to the City Corporation in 1197. The Crown regained control in 1857 after a legal tussle and assumed responsibility for the riverbed and banks (which it ceded to the Port of London Authority in 1908). This was a rare defeat for the City: kings, priests and nobles based at Westminster usually failed to gain control over the City authorities.

CROSS RIVER TRAFFIC

The City Corporation used its influence to prevent other bridges being built, and so it was quicker for most people crossing the river to use ferries or river taxis (wherries). These could be hailed by waiting at one of the stairs dotted up and down the river or by standing on the bank and shouting 'oars! oars!' or 'scullers!'. It has been claimed that in the sixteenth century, 40,000 watermen worked between Gravesend and Windsor. This number had halved by the early nineteenth century and by 1850 only 4000 or so remained in the metropolitan London area. The watermen opposed bridge building as it clearly affected their trade, and often when bridges were built it was only after the watermen had been compensated for loss of income. Like their black-cab brethren today they were wary of any perceived threat to their income, whether from enclosed docks, steamships or sedan chairs. Unlike their twenty-first-century counterparts, however, they were always happy to go over the water.

Bridges may have been bad for the watermen but others saw them as a source of greater prosperity because they would speed up and increase commerce, and by the eighteenth century powerful forces militated for more crossings. These included large landowners who possessed property upriver, as well as traders and businessmen who wished to open up direct links between the growing populations north and south of the Thames. Parliament was also interested in removing the monopoly of Old London Bridge as a means of restricting the power of the quasi-autonomous City Corporation. A second bridge linking Fulham with Putney was finally built in 1729, and when speculators in Surrey and Kent again pushed for a crossing at Westminster, the owners of the Fulham Bridge, who were doing very nicely from the tolls, now predictably joined the opposition. Westminster Bridge was eventually completed in 1750 with the aid of monies raised from five lotteries and a government grant. The City authorities responded by clearing London Bridge of houses to speed up access, and later they opened a new bridge at Blackfriars.

Other, privately financed, bridges rapidly followed, many on the sites of the old fords where the river was narrower, and approached by an existing network of roads. Bridges were considered good investments because tolls could be charged and they were a more reliable means of crossing the river than boats. The greatest period of bridge building coincided with London area's most rapid population growth, from half a million in 1750 to six million in 1900. Starting with Westminster thirteen new road or foot crossings were added, starting at Blackfriars (1769), Battersea (1771), Vauxhall (1816), Waterloo (1817), Southwark (1819), Hammersmith (1827) and Hungerford (1845).

Later crossings came at Chelsea (1858), Lambeth (1861), Albert Bridge (1871), Wandsworth (1873) and finally Tower Bridge (1894). The inner London railway bridges all came in a burst between 1859 and 1869, with Putney added in 1889.

In the 1880s the Metropolitan Board of Works bought out all the private bridge companies and made the crossings toll-free. The building of bridges, along with the embankments and creation of the London-wide sewerage system, not only shaped the look of the city and behaviour of its citizens but also had a fundamental impact on the city's government. The Board was also responsible for the building of new thoroughfares, slum clearance and park landscaping, and became effectively the government of London. Had it not been successful in these ventures, the enemies of a central government for London might have won out and the London County Council might not have been created in 1889.

CROSS RIVER TRAFFIC

A third of London's population lived east of London Bridge but the Thames there was a working dock area and bridges presented an obstacle, so other solutions were adopted. The Metropolitan Board of Works introduced the free Woolwich ferry in the 1880s but before that several tunnels had been dug. In 1798 Ralph Dodd tried to build one under the Thames between Gravesend and Tilbury, but this was halted when he ran into quicksand and out of finance. In 1808 Robert Vazie and Richard Trevithick got within forty metres of Limehouse shore when the Thames broke through. They cleared the workings, but when the same thing happened again the project was abandoned. In 1843 Isambard Kingdom Brunel completed the tunnel (between Rotherhithe and Wapping) started by his father, Marc Brunel, in 1825.

In 1869 the Brunels' 400-metre passage was taken over by the East London Railway and today the East London Line runs through it. A year later James Henry Greathead opened a new tunnel from Tower Hill to Vine Street which initially had a cable car service but was soon converted into a foot tunnel that attracted a million customers a year before closing in 1894. It has since been used for water mains. There are currently twenty tunnels under the Thames downriver of Hammersmith: two are foot passages at Greenwich and Woolwich, while the rest act as conduits for water, electricity, gas, cars and trains. Below the tube lines at Hungerford and Blackfriars are two deep tunnels built by the Post Office for their own rail network to transport mail across London. These are now part of the defence network for London in the event of an attack – an interesting echo of one of London Bridge's early functions, to defend the city and protect its boundaries.

These city boundaries were last expanded in 1965, when the Greater London Council took in more boroughs including Newham, Hounslow and Richmond. The GLC then became the Greater London Authority, which today, via the local borough councils, is responsible for the lighting and maintenance of the western bridges, paid for through taxation. The foot and road bridges to the east of Waterloo, meanwhile, bear the mark of the Bridge House Estates (designed by the seventeenth-century surveyor William Leybourn). This 900-year-old charity, charged with looking after a bridge demolished over 150 years ago, still has power, influence and a positive effect on the lives of millions of Londoners. The Estates built and are responsible for the upkeep of Blackfriars, Southwark, Tower and London bridges. They also maintain the Millennium Bridge. This latest crossing came at a time when London's population was increasing again after nearly half a century of decline from 1939 to the 1980s. There has been an annual increase of 50,000 people a year over the past decade and it is projected that by 2020 the population of the city will top the eight-million mark again. Additional bridges are planned for the east as the world's first metropolis starts to refocus on the Thames, its lifeblood and very reason for existence.

PASSPORT TO PIMLICO

Vauxhall Bridge

In 1998 the Thames Archaeological Survey found the remains of an oak crossing built 3500 years ago not far from where the river Effra empties into the Thames at Vauxhall. Two parallel lines of large oak posts (forty centimetres across) led into the river; they were spaced about five metres apart, which implies that they were part of a bridge rather than anything less substantial, such as a jetty. The bridge may not have crossed the whole river but instead maybe formed a link to a long-disappeared island in what would have been a much broader Thames. When samples of the bridge were tested evidence was found of an alga that only lives in estuaries, suggesting that the tide ran as far as Vauxhall and that this may have been the point where the tide turned.

The Thames river valley has been inhabited by humanoids for a quarter of a million years, although homo sapiens did not arrive until much more recently. There is evidence of settlements at Bermondsey, Battersea, Wandsworth and Vauxhall thousands of years before Christ and, more importantly for the purposes of this book, before the Roman city of Londinium was founded. But in one sense this is the original 'London' bridge site. It was not until 1816, however, that the first iron bridge to span London's Thames was completed at Vauxhall. Lord Dundas, representing the Prince Regent, laid the first stone on the Middlesex side in 1811 for what was at one point to be called the Regent's Bridge. Work on the Surrey shore started two years later in 1813, when Prince Charles of Brunswick laid the first stone there. John Rennie originally planned a blue sandstone bridge with seven arches, but as finances were tight he suggested a new, lighter design of eleven cast-iron arches. This was not accepted either, and after some delay, a Mr J. Grellier was commissioned to build a nine-arched iron bridge to a design by Sir Samuel Bentham. The Bentham plans were abandoned because of doubts over the quality of the work and fears of conservation bodies that the proposed bridge would adversely affect the flow of the Thames.

Eventually it was James Walker who built the 246.5-metre long (eleven-metre wide) granite-faced, cast-iron structure with eight stout stone piers. Vauxhall Bridge was regarded as a decent and practical addition to London's collection of bridges but nothing particularly special. It was light and airy in appearance, with nine equal arches, and it cost a halfpenny for a pedestrian to cross. The Vauxhall Bridge Company hoped to make a good income out of people going to and from the Pleasure Gardens that once stood where Spring Gardens and Vauxhall Cross are today. It was also planned that the bridge would lead to the development of the area south of the river by linking it to the well-populated spots to the north and establishing a land route from Hyde Park Corner to Greenwich. The area immediately north of the bridge, described by Charles Dickens in *David Copperfield* as 'a melancholy waste of coarse grass and rank reeds straggled over all the marshy land in the vicinity', was in need of some development too.

On the north bank was the vast Millbank Penitentiary, which occupied the current sites of both the Tate Britain and the Chelsea College of Art. The prison was built to a revolutionary new design, based on hexagonal blocks, that allowed prisoners to be more effectively monitored. It failed in its stated aim of reforming the inmates, and in its final days was primarily used to house convicts in chains before they were deported to Australia and elsewhere. The prison was demolished in 1867, and a stone mooring post is all that remains of it. A statue entitled *Locking Piece*, by Henry Moore, whose studio was in Vauxhall, stands near the post, in a small public garden by the river. The prison is also commemorated, allegedly, in the hexagonal architecture of River House, adjacent to the bridge.

All the delays to the construction of Vauxhall Bridge had raised the costs of the crossing, and it was not until the mid-nineteenth century, when the population of south London increased, that the Company saw any decent return on its investment. Despite the best precautions of James Walker, who dug deep into London clay to anchor the pilings and built heavily protected abutments and piers, tidal scour still weakened the structure and tons of material had to be dropped around the piers to bolster them. Plans to replace two central piers in 1881 with a single one to help the flow of river traffic were abandoned on grounds of cost. Constructing an entirely new bridge proved cheaper than repairing the old one, particularly when divers found that virtually the only thing holding it up was the arches. An interim wooden structure was thrown up until the new Vauxhall Bridge was completed in 1906; this was the first London bridge to carry trams.

Again there was some disagreement over the Vauxhall's design. Sir Maurice Fitzmaurice suggested five concrete and granite arches, but the actual bridge (built by Sir Alexander Binnie) has five steel arches supported on granite piers. It is 246 metres long and twenty-four metres wide, and is painted mostly in burgundy and orange with a blue and white trim. Each pier is topped by an impressive statue of a woman, the height of a double decker bus, sculpted by F.W. Pomeroy (upstream) and A. Drury (downstream). Both sculptors' statues are similar in style, dressed in gowns, and made of bronze. Pomeroy's figures represent various trades: a woman holding a pot for pottery; an engineering lady cradling a very detailed steam engine and a mallet. Architecture is next, holding a half-metre model of St Paul's Cathedral in her left hand and callipers in her right. Finally the figure closest to the north-west bank clasps a shepherd's crook and a sheaf of corn to represent agriculture. The downstream-side sculptures represent fine art, education, local government and science. The figures are well lit at night and when seen from the river can be quite alarming for those who remember the bronze statue of Talos coming to life in the film version of *Jason and the Argonauts*.

The bridge's two-metre-high ornamental iron railings were removed when the roadway was widened in the 1970s (you can still see where these railings joined the granite abutments): the Greater London Council had argued that the bridge needed to lose some weight and that the railings had to go. Lambeth and Westminster councils suggested lightweight replicas of the originals, but they were overruled. The low-level box girder replacements were not popular at the time but they do improve views from the bridge, which take in the Battersea Power Station and the

London Eye. The bridge is set on such a sharp bend in the river that these landmarks appear to move from the south to the north bank as you cross over the middle of bridge heading north.

The road over the bridge provides easy access to Victoria Station through the charming area of Pimlico, whose name appears to have a connection with Walter Raleigh's former US colony on Roanoke Island in the Pamlico Sound in present-day North Carolina, making it the first Native American name to be introduced into Britain. The earliest reference to Pimlico dates from 1626, although a map of 1630 uses the title 'Pimplico'. Vauxhall, originally Faukeshall, from Falkes de Breaute who had a hall there in 1233, is an older settlement, but by the time the bridge was built Pimlico was much more comprehensively developed. Pimlico gives its name to one of the finest of the Ealing comedy films, *Passport to Pimlico*, although this was actually shot over the water in Lambeth. It would be lovely to believe that the use of purplish-red paint on the bridge's ironwork is a subtle tribute to the film in which the people of Pimlico discover that by right of an ancient charter the area is actually part of Burgundy, not England. This prompts them to declare independence from the UK and ally themselves to the Duke of Burgundy. Like the film, Pimlico has a slightly subdued charm and gentility, whereas other films featuring Vauxhall Bridge, such as *Alfie* and *Theatre of Blood*, have less of either of those qualities.

PASSPORT TO PIMLICO

Vauxhall Bridge

Vauxhall has a long association with invention and engineering. In the 1630s an experimental workshop was set up in the area by Edward Somerset, Marquis of Worcester, a prolific inventor with interests in ciphers, locks and military engineering, who has been credited with the first practical use of steam power. In later years the area gave its name to the Vauxhall car, which was developed there. More strangely it gave the Russians their word for railway station, 'vokzal'. There are several theories as to how this came about but the most likely is that Vauxhall had become synonymous with the Pleasure Gardens, and the first Russian railway line went from St Petersburg to a nearby pleasure garden at Pavlovsk.

It's hard for the modern visitor to believe that for almost a hundred years from 1661, one of London's, indeed the world's, most famous parks was in Vauxhall. In 1820, 61,000 people crossed the bridge to attend a masked ball there. The gardens had many themed sections, including a re-creation of ancient Rome, fairground attractions and the first commercial use of a balloon ride as a viewing platform from which to overlook the city. The gardens were the talk of Europe; royalty and nobility from the continent and just about every British celebrity visited them. Boswell remarked:

> Vauxhall Gardens is peculiarly adapted to the taste of the English nation; there being a mixture of curious show, gay exhibition, music, vocal and instrumental, not too refined for the general ear, for all which only a shilling is paid.

By the second decade of the nineteenth century Vauxhall Pleasure Gardens lost their fashionable status and the crowd became rougher, as did the working girls who loitered in the walkways. In response, the admission price was lowered, attracting more ne'er-do-wells and even fewer of the gentlefolk who had first popularized the pleasure gardens. In a bid to reverse this trend the price was raised again and more upmarket attractions were put in, including a fountain containing a statue of Neptune driving five seahorses snorting streams of water from their nostrils. The gardens' fortunes revived slightly, but the 'Great Stink' of 1858 made any Thameside entertainment difficult to sell and the gardens closed in 1859. However, there was alternative river-based entertainment that started from the bridge in the mid nineteenth century: the sight of Thomas Barry, who was famous for sailing between Westminster and Vauxhall bridges in a tub towed by four geese.

Vauxhall carries on its tradition of hedonism into the twenty-first century, with more nightclubs in the area than just about anywhere else in London. The more specialist of these continue to draw people from across Europe and from all sections of society. These clubs are not all easy to spot but they are there, underground, hidden away, secretive, which fits rather well with the fact that the MI6 (Secret Service) headquarters is next to the bridge, on the Albert Embankment. The embankment has been around since 1869 and was part of a scheme to protect the area from flooding, as Vauxhall was once a marsh and flood plain of both the Thames and the Effra. In the late twentieth and twenty-first centuries large-scale housing projects were built on the embankment itself – but it is the bridge that is responsible for the layout of the neighbourhood, as settlement started along the arterial roads and then larger estates were laid out inland. Much of the land is owned by the Duchy of Cornwall, and the names of the streets along Kennington Lane have a number of Cornish themes and associations.

Underneath the MI6 building is the overflow pipe for the river Effra (which is just big enough to launch a mini-submarine from), and the Effra itself enters the Thames from beneath the St George's development just upstream of the bridge. Almost opposite this on the north bank is the entry point into the Thames of the Tyburn river, which is commemorated by a plaque detailing that river's course. This confluence of rivers and an ancient tidal turning point would have made this a particularly sacred site for the Bronze Age tribes for whom each river was a separate spirit, and the tide has associations with the moon goddess. More practically, rivers were a provider of food and wealth, but also dangerous and unpredictable forces that needed to be appeased or evoked.

In the twenty-first century people still conduct religious rites on the banks of the Thames. On 21 September 2001, the headless torso of a boy was found floating near Tower Bridge. As part of the investigation into this horrific discovery the police found many sites along the river where voodoo offerings, of a less grotesque nature, had been made. In 2004 a Hindu shrine was found near Chelsea Bridge, and some Christian groups baptize people in the Thames. There is a tradition of making sacrifices to bridges in order to ensure they remain in place too. In London these range from the Roman coins and statues found near Old London Bridge, to the casting of oil and corn on the waters at Hammersmith in the eighteenth century, to the burial of objects under Waterloo in the twentieth century.

From the Bronze Age to the space age, people have lived around a crossing over the Thames at Vauxhall. Alongside the ancient bridge were artefacts (axes among other things) made as offerings to the spirits of the river. Today the bridge's bronze statues pay tribute to more secular achievements, but a representation of the god of the river can be found just upriver of Vauxhall Bridge on the southern walkway. As the Romans prayed to Father Tiber, so Britons prayed to the Thames, which is lined from its source by sacred sites and groves. Old Father Thames is frequently portrayed as a bearded man, like the traditional guardians of the City (Gog and Magog), and his name acknowledges that the river itself is father to the settlements on its banks.

LONDON BRIDGE

MOVE ON UP

London Bridge

'London Bridge is Falling Down' is a popular childhood rhyme known across the world. Rather than being a critique of British talent when it comes to engineering projects, it is a simple comment that over the centuries London Bridge has been in need of almost constant repair. Not surprising, since there has been a bridge on roughly the same spot since Roman times. Some London bridges are better chronicled than others; for one crossing, the only proof for its existence is a reference to it in a witchcraft trial as being a place of witches' execution. This Saxon crossing was one of several wooden bridges destroyed in turn by Vikings (1014), by storm (1091), by fire (1136); the last, completed in 1163, was demolished to make way for the first stone bridge in the early thirteenth century. A later bridge distinguished itself by being exported in 10,246 pieces as a 'large antique', whereas the current crossing barely has any distinguishing features at all. What they have in common though is that they all sucked people and goods into the City, acted as a magnet for the desperate, the criminal and the ambitious, and funnelled trade towards London. The historian R. Wheeler takes this a step further and suggests that London was a parasite living off the bridge.

CROSS RIVER TRAFFIC

The Roman city of Londinium was established in the first century AD around the high ground of Cornhill and Ludgate Hill above the marshy flood plain to the north of the Thames. The first Roman bridge was a temporary military one replaced in c. AD 120 by a sturdier structure, at a site which held for much of the following 2000 years the only bridge in London. The Romans' choice of crossing place would have been based on the narrowness of the channel, the firmness of the ground on both sides of the water and the fact that the river was still tidal at this point, thus permitting easy, speedy access to and from the sea. The bridge also allowed the rapid deployment of legionnaires to the south and provided a good defensive feature for the city.

London Bridge had a dual role of blocking the river to shield the city from assault by ships, and enabling defenders to hold the southern shore. This didn't always work out: King Olav towed away a Saxon bridge in 1014 by attaching his ships to it, and a few years later, King Cnut sailed around it using a channel to the south. The first stone bridge, commonly referred to as Old London Bridge, had fortified gates beneath the towers on the entrances. There was a drawbridge, between the sixth and seventh piers coming from the south, that could be raised to make the city even harder to attack. As a warning to those even considering violence, the heads of executed traitors were displayed on the bridge approach, a fact that is commemorated by a giant white spike on the current crossing. The idea of London Bridge as a defensive feature ended for a time in 1823, when work began on John Rennie's bridge. It was the first bridge not to have a drawbridge or fortified approach; it was also the first bridge not to be built on the exact site of the Roman crossing. However today, on the northern approach road, there is a police checkpoint that monitors traffic and guards the City from terrorist attacks.

Bridge building was considered holy work. London's citizens were encouraged to make offerings of land and money 'to God and the Bridge' – and Old London Bridge demanded a good deal of both. It was completed in 1209 and took thirty-three years to build, under the direction of Peter de Colechurch. It was 274 metres long, with a four-metre-wide roadway and buildings either side, and it was supported by nineteen piers. In order to protect the piers and break the flow of the Thames these were surrounded by starlings (boat-shaped pilings). This first stone bridge was a defensive fixture but also part of the City itself, incorporating shops, houses and a chapel. The bridge was the scene of lavish celebrations: people lived, traded, worshipped and were even buried on it for centuries, until increasing congestion (and competition from other crossings) led to it being cleared of houses by 1758, in an attempt to speed up crossing times. At the same time the bridge was widened, one of the piers was knocked out and replaced with a larger span to ease the flow of the river and river traffic. George Dance (the elder), ironically the last man to build new houses on the bridge, was given the job of demolishing the buildings. A brief period followed (up to 1782) when tolls were charged on the cleared crossing, but when these were abolished the bridge became just as congested as before. In 1800 a committee was set up to suggest improvements to the bridge, but gradually opinion moved in favour of a new crossing altogether, as tidal scouring had weakened the bridge's supports much more than expected. The severe frosts of 1813 to 1814 settled the matter. After the iced Thames and the ferocious waters unleashed by the thaw caused further damage to the bridge's abutments, submissions were invited for the design of a new bridge.

The replacement was a five-arched granite bridge designed by John Rennie and completed by his son Sir John Rennie in 1831. Although Rennie's bridge was a beauty, classical in appearance, it was nowhere near as dramatic as the 183-metre suspension bridge proposed by Thomas Telford or the spectacular redevelopment plans of George Dance (the younger), who envisaged huge piazzas at either end of parallel viaducts, with a massive drawbridge for large ships. Dance didn't put forward his plans in the 1820 competition but Telford did, as did Robert Mylne and C. Fowler. Fowler actually won the competition, only for Parliament to overrule the committee and select the Rennie design because he was considered a more reliable and trusted engineer.

Old London Bridge remained in place until 1832, nine years after work began on the new crossing and a whole network of approach roads. Rennie's bridge was 306 metres long, seventeen metres wide and thirty metres upriver of the old crossing point. It had nicely curved cutwaters and robust buttresses that provided an elegant and stately approach to London and was very heavily used, with 22,000 vehicles and 110,000 pedestrians crossing every day by the late nineteenth century. Charles Dickens Jr noted in his *Dictionary of London*:

> In order to facilitate traffic, police-constables are stationed along the middle of the roadway, and all vehicles travelling at a walking pace only are com-

> pelled to keep close to the curb. There are still, however, frequent blocks, and the bridge should be avoided as much as possible, especially between 9 and 10 a.m. and 4 and 6 p.m.

He went on to add that 'seen from the river, it is the handsomest bridge in London'. It was easier to admire from the river than Old London Bridge because the broader spans of the four piers meant that there was better access for ships. Unfortunately this relatively unrestricted access also provided the tide with an opportunity denied it for centuries to really explore inland and rampage a long way upriver. In the past, Old London Bridge had held back the freshwater behind it and arrested the incoming tide, while the more flexible broader banks meant that the tide would exhaust itself before reaching too far upstream.

Removing the bridge and embanking the river meant that a faster-flowing Thames' tidal reach only stopped at the artificial barrier of Teddington Lock, and there was a fierce debate about the likely consequences. The chief fears were that the unleashed current would undermine the existing wharves and flood large (and by now) heavily populated parts of the city. H.G. Englefield predicted many of these in his 1830 study but, although he highlighted sewage as an issue, he did not anticipate it bobbing up and down through central London as the tide rose and fell. The destruction of Old London Bridge did however have the positive effect of removing the dangerous rapids caused by the closely positioned piers. Braving the turbulent waters beneath Old London Bridge was known as 'shooting the bridge' because of the speeds with which boats were hurled through by pressure of the river and the drop of up to two metres either side. It was a dangerous and often fatal activity with thousands of people dying over the centuries, including in 1290 a shipload of Jews who were being expelled from England by Edward I. It is alleged that their cries can still be heard on quiet nights from the embankment near Customs House.

The problem of tidal scour was not immediately apparent as it took Sir John Rennie some time to clear away the rubble of the old bridge. Much of that debris was reused on other building projects but two alcoves can still be found intact, at Guy's Hospital and at Victoria Park, Hackney. Over the centuries offerings had been made to the bridge and valuables dropped onto the piers, so when the bridge was removed Roman coins and other articles were found at the site. Coins were also later found at Putney and other places across London where material from the old crossing was recycled to build embankments and paths. Perhaps the strangest collection of fragments are in an armchair in the Fishmongers Hall, which is made up of wooden pilings from Old London Bridge, Rennie's London Bridge, and the first Westminster and Blackfriars crossings.

As Rennie's bridge became increasingly crowded, plans were made in the 1880s for a bridge on the site of the old crossing, but this never materialized; nor did the 'Metropolitan Bridge' proposed for the same spot in the 1980s. However, a bizarre vestige of the old bridge will remain for another eight decades or so, in the form of stocks and shares. In 1581 Peter Morris bought a 500-year lease on waterwheels fixed to the old bridge. These were sold to Richard Soame in 1701 and, even when water was no longer allowed to be drawn from the Thames, his descendants were given shares in the New River Company, which became the Metropolitan Water Board and is now Thames Water. The shares still carry dividends until the lease runs out in 2082.

Despite being widened in 1902, by the 1950s even Rennie's bridge was unable to manage the tide or the increased traffic flows of the twentieth century, and a decision was made to replace it. But it was famously preserved after being bought in 1968 by the McCulloch Oil Corporation of America, which came top of many bids. They shipped it across the Atlantic and had it re-erected, piece by piece, over a bit of Lake Havasu in Arizona – originally over dry land, with the water channelled through later. This removal was a wonderful, if somewhat eccentric, feat of organization and engineering, and the Americans even got an unexpected bonus when people reported seeing four ghostly women in Victorian costume on the bridge during the dedication ceremony. Other bits of this crossing remain in London, including the stairs on the south side (named Nancy's Steps after a murder scene from *Oliver Twist*) and some parts that were removed to an ornamental pond in Kew Gardens.

The current crossing was designed by Mott, Hay and Anderson and officially opened in 1973. Between 1967 and 1972 Rennie's bridge was dismantled as the latest model was being constructed. It's an odd coincidence but the price of the new three-span pre-stressed concrete bridge, at just over £4 million, was more or less the same as the cost of buying, transporting and re-erecting the Rennie bridge in the US. The current bridge is almost twice as wide as Rennie's, at thirty-two metres, and has four traffic lanes and two footpaths. The concrete has been given a granite finish and the steel handrails complete the slick, but soulless, appearance. Perhaps in an attempt to appear cosier there is a heating system below the surface of the roadway and pavements to prevent freezing.

It is a shame that something as famous as London Bridge should now resemble nothing more than a suburban flyover. Beneath it on the north bank there is a good view of the rather elegant piers and delicate spans, but these only flatter to deceive, as the municipal product that rests on them fails to deliver any sense of grandeur. The network of streets, alleys and stairs underneath it is confusing, but at least provide a sense of Old London. They also offer evidence of the failure to more comprehensively restructure the approach roads in the nineteenth century. Although Borough High Street was redirected on the southern side, the planned piazzas for the north never materialized and the buildings around the bridge, which range from sturdy Victorian to modern office block, are not what you might expect from the approach to arguably the best-known bridge in the world.

London Bridge's chief function is for the bondsmen, and -women, to cross from London Bridge Station (the first train station completed in London) and other points south to their places of work on the north bank. This makes for an interesting north–south tidal effect on London Bridge, in contrast to the east–west one beneath it. It is extremely difficult to cross from the north to the south during rush hour in the morning and from south to north in the evening. Tourists and others moving against the flow find themselves buffeted by the besuited hordes, like salmon braving the currents. The dullness of city commuting is a staple of British comedy, poetry, popular song and literature, but there are cheerier sights to be seen on the bridge, such as the annual walking of a flock of sheep into the City – a right open to all Freemen of the City of London, and a nod to the wool tax that helped pay for the original stone bridge. The more cynically minded though might see this as the perfect metaphor for London Bridge: lambs to the slaughter or just another mindless flock heading to the trading floors.

London Bridge, along with Westminster and Waterloo, is one of London's busiest pedestrian crossings and, despite the introduction of congestion charging, is still crossed by over 30,000 vehicles a day. This is less than Tower, Blackfriars and Waterloo (circa 40,000) and well below the nearly 60,000 of Wandsworth and Putney, but only Waterloo carries anything near the number of buses. Since the congestion charge there has been a large shift towards public transport in central London and a less dramatic one towards bicycles and other forms of two-wheeled transport. In total 320,000 non-motorists cross the inner London bridges every workday.

This flood of people across London Bridge, and indeed all the bridges, is more diverse racially than ever, but since Roman times there has always been a decent mix of people. A quarter of seventeenth-century London's population were migrants (mostly from the UK, but many from overseas), and the Victorian city that grew by five million in a century sucked in people from across the globe. A third of inner London's population is now ethnic minority, with British-born Asians and African people joining the workforce in greater numbers alongside a recent influx of non-native employees. Women have always made up part of the commuting workforce but now they are more likely to be equal(ish) partners on the trading floor not, as Arthur Munby noted in his 1861 diary, doing all the menial jobs:

> London Bridge, more than any place I know here, seems to be the great thoroughfare for young working women and girls. One meets them at every step: young women carrying large bundles of umbrella-frames home to be covered; young women carrying wooden cages full of hats, which yet want the silk and the binding; costergirls, often dirty and sordid, going to fill their empty baskets.

There are more bridges for the women (and men) of south London to cross today and greater opportunities for them to fill their baskets, but for centuries the City of London, and the innkeepers of Southwark, jealously guarded the monopoly it had on downriver Thames crossings. Other bridges were prevented from being built, allowing the City of London to prosper, but when the rival Thames bridges were erected they allowed a greater London to expand from the original walled city. This bridge is where it started though, and millions of people over hundreds of years have crossed a London bridge, following a tradition of individuals making their fortune in the city that goes right back to the early Roman settlers. They are attracted by the opportunities, by the affluence, by the power of the city by the Thames. Those Romans who built a trading place on a tidal river and erected a bridge over it could have had no idea that their actions, in an outlying province, laid the foundations for a city that would eclipse Rome itself, and become the centre of an empire larger, more powerful, wealthy and diverse than they could ever imagine.

LONDON CALLING

Putney Bridge

Putney Bridge links Putney in the south to Fulham in the north, two long-established settlements which were swallowed by the metropolis during its huge growth in the nineteenth century. Putney gets its name from an Anglo-Saxon chief, Putta, and simply means 'Putta's Landing', and is an area that Sir Arthur Conan Doyle once referred to as the cultural desert of south London. This is pretty scathing considering some of the competition, as well as unfair because former residents include Oliver Cromwell and Edward Gibbon. Furthermore nowhere that has a shrine to Marc Bolan (on nearby Barnes Common) could ever be described as being entirely devoid of culture.

Fulham derives from the settlement of Fulla on a bend in the river and started life as a fishing village. It later became a semi-fashionable retreat for the wealthy, then from the nineteenth century developed into a working-class suburb that, up until the 1960s at least, was rough enough to have its own criminal syndicate. It wasn't quite Bethnal Green or the Elephant and Castle, but a contender nonetheless in the hierarchy of London firms. Fulham Palace is the traditional home of the bishops of London, who were unique in being the only people not to have to pay the toll on what was originally known as Fulham Bridge, a wooden crossing completed in 1729.

There are records of a ford, and ferries from Putney are mentioned as far back as the *Domesday Book*, but the first proposal for a bridge came in 1671 and was sponsored by John Dwight. Dwight lost out, however, to those opposed to the crossing, who argued that a bridge at Putney would result in the annihilation of the City of London by shifting trade west. They further contended that such a crossing would bring about the destruction of the Thames by blocking the river which in turn would cause flooding and disrupt fishing. The area might have remained bridge-free had the journey to London of Prime Minister Sir Robert Walpole not been interrupted fifty years later: all the local ferrymen had refused to leave the Swan pub, where they were having a nice drink and a chat, to help the PM across. Walpole petitioned Parliament for a crossing in the 1720s and in 1726 an Act of Parliament authorized the construction of a bridge, but only if the ferry owners were fully compensated, in beer presumably. It would seem that the fears of the anti-bridge lobby were misplaced as, over 250 years on, the City of London is now more opulent than ever and the Thames, admittedly after a period of stagnation, is thriving.

The bridge started (and ran through) the churchyard of St Mary's on the

Putney side, though builders were careful not to disturb the gravestones, and ended near what is now Fulham bus station. St Mary's was the location of the Putney Debates held in 1647, where the self-styled Levellers proposed a whole spectrum of revolutionary ideas just after the English Civil War. These included voting rights for all adult males, trial by jury, complete religious freedom, an end to the censorship of books and newspapers, and the abolition of the monarchy. On the north side in Bishop's Park is a tribute to the International Brigade of volunteers who went to fight in the Spanish Civil War against Franco's fascists for similar principles (with equally dismal results) in the twentieth century. Bishop's Park is also the setting for another struggle between good and evil (nothing to do with nearby Fulham FC): it was used in the film *The Omen*.

Old Fulham Bridge was a curved timber affair supported by twenty-six arches, built by Thomas Phillips to a design by Sir Joseph Acworth. It was attractive but required considerable maintenance, not least because ships kept colliding into it as the structure was so low in the water. In 1870 a barge damaged three central sections and the following year two piers were removed and replaced by a twenty-one-metre iron girder. In 1879 the Metropolitan Board of Works purchased the bridge and decided that it was time to build a new one. The twenty-five piers were close together and had presented a major obstacle to navigation by the watermen. As did early feminist and writer Mary Wollstonecraft who was rescued by a passing boat when she threw herself off the bridge in a rather long-winded attempt at suicide. Having failed to leap from Battersea Bridge because it was too crowded, she then took a boat upriver, from which she intended to jump, but once again she chickened out; she ended up in Putney. There she finally dived into the river but nearly hit a boat and was easily picked out of the Thames.

The current Putney Bridge was completed by Sir Joseph Bazalgette in 1886 after four years' work. This is Bazalgette's first London bridge and it is certainly the most restrained, with the cool passionless sweeps reminiscent of crossings in moneyed market towns such as Perth. It is slightly upstream of the first bridge and follows the line of the former Chelsea waterworks aqueduct (whose pipes ran under its footpath). An operation in 1933 to broaden it to twenty-five metres wide now allows generous space for pedestrians to admire the almost rural views upriver as the Thames curves away towards Hammersmith. The approach roads directly feed from Putney High Street, and midway across is a marker of the boundary between Fulham and Putney. Interestingly the only person compensated when the new bridge was erected was the landlord of The Eight Bells pub, whose trade was compromised by the tavern no longer being at the end of the bridge.

There is a slightly Palladian influence on this chaste five-span crossing and, until modern buildings overshadowed it, the bridge blended well with its rather refined surroundings. It is a solid, straight structure of quiet dignity and beauty built out of Cornish granite with concrete supports. Its four piers have far less of an effect on the river's flow than the previous bridge, and although one enthusiastic commentator described it as 'a rainbow in stone', it is less ornate and more traditional than Bazalgette's other Thames crossings. The only decorative features are the eight sets of trident lights on the balustrades and the single ones at each approach, which are etched at their bases with a series of heraldic designs. These crests include one that looks like a stumbling donkey, though apparently is a horse *ployé*, symbolizing readiness to serve authority. The lack of flourish, the traditional structure and the materials used give the impression that the crossing has been there much longer than a century and a quarter.

There is another crossing nearer the site of the old Fulham Bridge, designed by William Jacomb and completed in 1889. This is a wrought-iron lattice girder bridge placed on two piers and four sets of iron girders all painted a light green. There are eight spans to the bridge, including two over land on the south side and one on the north. This robust and practical structure carries the District Line towards Southfields. At one time it was referred to as 'the iron bridge', despite the abutments of brick with Portland Stone dressings on either shore. Downriver, almost it seems attached as an afterthought, is perhaps London's loneliest pedestrian route, linking the bus station on the north side to East Putney Railway Station. Its vaguely industrial feel and proximity to the track should appeal to those who miss the old Hungerford Bridge: the ornate top of the pier, back and front, can be viewed across the railway track through a mess of wire and there is that same beautiful shake of the bridge caused by a passing train. There is also the knowledge that you are only a flimsy municipal handrail away from the river seven metres below.

With its ornamental cast-iron parapet, it's prettier, straighter, better drained and less threatening than Hungerford was, even if the graffiti-spattered and dark approaches from the Fulham end are a little menacing. At the Putney end, in contrast, is a classic metroland of semi-detached houses – not at all disturbing, except for the fact that some are built over a plague burial pit. There are good views off to the east of the old-style suburbia merging into the newly built waterfront developments, providing an interesting contrast between the old and new Thameside. Also downstream, between Wandsworth and Hurlingham Park, was the spot in 1642 where a bridge of boats was built by the Parliamentarian Army. That crossing was a short-lived military necessity – or just possibly a means by which Putney resident Oliver Cromwell could nip across to the north bank to argue with the bishops.

The twenty-first-century river is alive with boats as well, though not enough to walk over it on, and since 1845 the university boat race between Oxford and Cambridge has started on the Putney embankment. It's strange that this race is referred to as *the* boat race yet involves people from out of town coming up for the day. Races of greater antiquity (the Doggett's Coat and Badge) or local interest (the Barge Race, Head of Rivers Race, or the annual contest between the River Police and London Fire Brigade) sadly do not merit the same attention.

Access to the river is simple in Putney as it is the first extended stretch in London where it is not walled off. This is good for the rowing clubs who can get their boats to the water easily and provides walkers (who don't mind mucking up their shoes) a decent stroll along the foreshore. It also exposes the area to the rapidly rising Thames during times of flood, heavy rain or high tides, and parking alongside the river can be risky even after only a moderate rainstorm. On 6 and 7 January 1928 things were more than moderately damp as a northerly gale raised water levels in the Thames estuary and several places in London, including Putney and Hammersmith, were flooded. There was severe destruction as roads up to half a mile inland were covered and basement flats filled with water. Further downriver, a twenty-five-metre section of the embankment near Lambeth Bridge collapsed, causing the death of fourteen people. Battersea, Westminster, Southwark and Greenwich were also hit and Temple Station underground was flooded while ships rode the waves at embankment parapet level along the Thames.

The 1928 tide was 1.8 metres above predicted levels, but high-water level at London Bridge has increased by about seventy-five centimetres each century due to a combination of the UK tilting towards Europe (at a rate of thirty centimetres per century) and the embankments, which between Putney and London Bridge are

5.41 metres high, and which act to constrain the river. The risk of another flood is exacerbated by rising sea levels due to global warming and river levels climbing more rapidly due to the larger area of concrete and tarmac built on the flood plain. London is particularly vulnerable if a low-level weather 'trough' moves across the deep Atlantic and is squeezed through the North Sea towards the Channel and Thames estuary.

In 1953 floods killed 3000 people along the Thames and galvanized the government to set up a committee headed by Lord Waverley (better known as John Anderson, inventor of the wartime bomb shelter), which in 1965 agreed to construct a compact, attractive, practical and environmentally sensitive barrier across the river. Nearly another two decades passed before Charles Draper's 520-metre Thames Barrier at Woolwich became operational in 1982. Since then it has been raised about three or four times a year (though fifteen times in 2001). It will now last until 2030, after which it will need to be raised an extra metre, to cope with a projected further rise in water levels.

The barrier should keep the communities of Fulham and Putney and others who live by the river safe, and allow them to continue to chatter about rising house prices or indeed the high-rise buildings overshadowing the bridge rather than rising floodwaters. It is peculiar that a barrier should be now protecting the city when one of the arguments against building Putney Bridge in the first place was that it would block the river and wreck both the Thames and London. Still if the barrier does fail all that rowing practice will come in handy and, as a sea trout was found at Fulham recently, the communities could always return to being fishing villages on a broader river.

SALE
SALE
T/SHIRTS
£5.99

KICK OVER THE STATUES

Westminster Bridge

At the north-east corner of Westminster Bridge a noticeboard displays the quaintly phrased bridge bylaws, drawn up in 1892 by the old London County Council. There are prohibitions against (in order) injuring the statues, committing a nuisance, attaching ropes, damaging the bridge itself, disobeying the weight limit and regulating the traffic without proper authority. Many of these (with the possible exception of the penultimate one) are flouted every time demonstrators attempt to 'take' the bridge and cross to the Mother of Parliaments in order to vent their feelings. The police have often had to close the bridge in order to protect the Palace of Westminster from the south. This is ironic, as one of the reasons a bridge was built here in the first place was to enable troops to be quickly dispatched into the seditious south (a nest of fanatics, according to eighteenth-century commentator John Webb). Some didn't bother to cross the bridge to make a political point. The Greater London Council (until its abolition in 1986) used its position in County Hall, on the south shore, to publicize causes very effectively by erecting massive placards clearly visible from Parliament.

Bridging the Thames at this spot was controversial and the debate lasted (on and off) for 200 years. The watermen and others argued as they had done at Putney that the bridge would interfere with currents on the river, disrupt trade with points west of London, interfere with shipping, affect fishing and lead to a deskilling of the river workforce that would result in the country's navy being understaffed in times of war. Other initially sustained objections were that a bridge would dam the Thames and increase sewage or, rather, inhibit the swift dispersal of sewage by blocking the river's flow. These last points would have been true if a crossing like Old London Bridge had been planned, with its close piers and huge starlings that acted like a dam.

Most proposals for Westminster Bridge sought to avoid these problems by building narrower piers that would cause less disruption to the river currents. But no matter how little the bridge interfered with the river's flow there were those who opposed the crossing on principle. In the seventeenth century the Archbishop of Canterbury feared the loss of trade for the church-owned horse ferry just upriver, and the City elders along with the Aldermen of Southwark were concerned about a draining away of trade from London and London Bridge to outlying settlements like Westminster. So the church and the City Corporation, along with the watermen, formed an unlikely, but powerful, alliance to oppose bridge building in central London.

In 1736 the then Earl of Pembroke and a supporting group obtained the passage of an Act for 'The Building of a Bridge across the River Thames from New Palace Yard in the City of Westminster to the opposite shore of the County of Surrey'. The first designs were for a stone bridge, which switched to timber, then half-and-half, then stone again. The eventual passing of the Act was seen as a victory for Parliament (and the City of Westminster) over the City of London and the church. When it was announced that the bridge would be funded by a lottery (i.e. gambling), these groups and others united to call it the 'bridge of fools'. The lottery that was supposed to fund the bridge gave away eighty-four per cent of the money, but still hoped to raise £100,000 (it actually raised £40,000): some of the biggest winners were a few families of watermen investing in the end of their own trade. Other lotteries (five in all) met with similar lack of interest, so in the end the government stumped up a grant, making Westminster, along with Chelsea, one of only two bridges originally funded by public money.

Work was started in 1739 but was disrupted by a frosty winter, followed by thawing floodwater that swept away some of the piles. Delivery of the stone required was frustrated by the Spanish navy who sank the coasters on their trips from Purbeck and Portland. On top of this, the British navy started press-ganging the shipmen who were bringing the stones for the bridge upriver, and barges frequently collided into the half-built structure, sinking a pier in 1748. There was also the Westminster earthquake in 1750, the year the bridge opened, but it managed to cope quite well with that.

A Swiss engineer, Charles Labelye, won the commission to build the bridge, after he came up with the innovative idea of using pre-built caissons to support it. These were huge boxes constructed onshore then floated into position and driven into the riverbed, using an early form of pile driver invented by a Mr Valoue. Water was then pumped out and the caissons were filled with stone, and the weight of the masonry piers constructed above was designed to fix them in place. This was meant to speed up the process by avoiding temporary piers being driven into the riverbed, then being replaced by permanent ones. Labelye won the contract for the bridge partly due to this original approach, and also because his design was deemed the least obstructive of the river's flow. Both he and fellow competitor Nicholas Hawksmoor were keen to avoid having to use giant starlings to support the bridge, as these would block the river. Another competitor though, the architect Batty Langley, was quite, well, batty about the final bridge and its absence of proper supports. He described it as act of 'unparalleled gross ignorance, madness or knavery'.

Others thought it beautiful and this first Westminster Bridge was praised as one of the most complete and elegant structures of its kind in the world. Until it fell apart seventy years later, it had fifteen semi-circular arches incrementally diminishing from the centre, resting on fourteen piers. On each side was a fine balustrade of stone, with semi-octagonal turrets at intervals to provide shelter for pedestrians. Canaletto drew and painted it twenty times between 1746 and 1756, and it was also the bridge about which Wordsworth rather excitedly wrote:

Earth has not anything to show more fair:
Dull would he be of soul who could pass by
A sight so touching in its majesty.

According to some, 'Upon Westminster Bridge' was the height of poetic licence: far from seeing the rising sun 'glittering in the smokeless air', Wordsworth's view had been quite gloomy on the day he crossed. Though it may not have been as opaque as later in the nineteenth century when Monet spoke of the sun suspended in the air over the bridges being 'no stronger than the still light of a lamp'. This was fine for Monet and the other French artists who flocked to London for the (lack of) light. As Mallamé said: 'Shrouded in mist, it is an incomparable city.'

The bridge (and indeed the city as a whole) was much darker in the past with or without the fog. Gas lamps replaced the old oil ones from 1802, but the light shed by these was still patchy and localized, and even when electric lighting became widespread in the twentieth century there were still great fogs. As late as the 1950s people drowned after stumbling into the Thames during particularly bad pea-soupers. The five three-branched lights on the current bridge are decorated with roses and fleurs-de-lis, and the bases depict clusters of houses with domed pillars rising above rooftops and representative of the view downstream.

There are other sights from the bridge that not everyone sees. For example, at midnight every 31 December a leaping shadow is reputed to hurl itself into the river just as the new year begins. This phantom man is believed to be the ghost of Jack the Ripper, re-enacting the taking of his own life. Prominent Ripper suspect Montague John Druitt did commit suicide in the Thames and his body was found on New Year's Eve in 1888. But this was near Chiswick several miles upriver, within drifting range of Hammersmith Bridge, from which a man uttering the claim 'I am Jack the Ripper' was reported to have leapt. More enigmatically, a small boat with three people travelling on its deck has been seen going under the Westminster, never to materialize from the other side.

Muggers and prostitutes once lurked in the alcoves at the side of the bridge, finding the dry, sheltered spots ideal for practising their professions. At one stage twelve night-watchmen were hired to guard travellers crossing the bridge after dark, which should put into perspective both today's fear of crime, and contemporary Westminster Bridge-based criminal acts involving the sale of bent cigarettes and overpriced London tat to tourists. James Boswell made the crossing before the guards were in place and recalled:

> I picked up a strong jolly young damsel and taking her under the arm I conducted her to Westminster Bridge, and there in amour complete did I engage up in this noble edifice. The whim of doing it there with the Thames rolling below us amused me much.

It isn't recorded whether Boswell's exertions contributed to the structural problems with the bridge's foundations. Although it stood for over 100 years, problems of subsidence were discovered in 1823 and there had been repeated concerns about the bridge's safety virtually since its completion. At each crisis point the leading engineers of the day were called in to advise. Telford, Cubitt, Rennie, Brunel and Page all offered their opinions and some designs for an alternative bridge. In 1826 all the piers were encased in cofferdams to allow repair work to take place, but as soon as this was completed, Parliament decided to build a completely new bridge. Then it changed its mind and decided on a stage-by-stage reconstruction of the old one to be overseen by James Walker. Walker compared the bridge to a patient that has 'been in the hands of the doctors since the day of its birth'. In 1846 after this work was complete Walker, along with George Rennie, Charles Barry and Thomas Page, put in a design for a replacement that Parliament accepted.

Thomas Page was selected to build the bridge but Sir Charles Barry was brought in as consulting architect to harmonize the bridge with his new Palace of Westminster. The bridge is slightly to the west of Labelye's, and Page was determined that his crossing would last. He built the bridge on a sturdy base with huge foundations for the piers, which themselves consist of twenty- to thirty-ton granite blocks for protection against the tide. The seven arches are each constructed of fifteen wrought-iron ribs. The bridge was completed in 1862. It is painted an attractive olive green, to match the colour of the benches in the House of Commons. On the ironwork of the bridge are etched a portcullis, the cross of St George, a thistle, a shield and a rose: symbols of the United Kingdom and Parliament itself. Charles Dickens Jr was unimpressed and wrote in his *Dictionary of London*:

> Westminster Bridge varies very much in appearance with the state of the tide. It is always rather a cardboardy-looking affair, but when the river is full, and the height of the structure reduced as much as possible, there is a certain grace about it. When, however, the water is low, and the flat arches are exposed at the full height of their long lanky piers, the effect is almost mean.

This is slightly unfair as, with the exception of some minor repairwork in 1924, the bridge has remained as Page left it. Westminster is the oldest London crossing still in use and remains one of London's busiest footbridges. The only way to see the bridge empty is to watch the film *28 Days Later*, which makes fine use of a bridge that crops up in a whole series of disaster movies from the superb *Day of the Triffids* to the woeful *Reign of Fire*.

Ineffectively guarding the southern side of the bridge from killer plants and dragons is the south bank Lion statue underneath which, within the plinth, is a room where the security guards who patrol the riverfront can enjoy a cup of tea. On the north bank is Thomas Thornycroft's statue to the Iceni Queen Boadicea (Boudicca), but there are no statues here to the great Victorians who helped build up the city and saved the river that Londoners had successfully killed off through their sewage and other waste.

In 1857 Sir Michael Faraday wrote to *The Times* detailing his own experiences on a river trip along the Thames, during which he tested the water (using white cards) for pollution.

> Near the bridges the feculence rolled up in clouds so dense that they were visible at the surface, even in water of this kind. The smell was very bad, and common to the whole of the water; it was the same as that which now comes up from the gully-holes in the streets; the whole river was for the time a real sewer.

There had been a public latrine on Old London Bridge that plopped directly into the Thames, providing boatmen with a fresh source of worry, and other sewage reached the river through tributaries, sewers and overflowing cesspits. Disturbingly, much of the population was taking their drinking water from the same source. Pipes drawing water out were sometimes adjacent to ones pouring sewage in, substantiating the often-repeated claim that every drop of Thames tap water had previously passed through several people. This situation was partly addressed in 1852 by the Metropolitan Water Act, which prohibited removing water downriver of Teddington and decreed that all water supplies be filtered. These measures were prompted by outbreaks of cholera in 1832 and 1849. The latter epidemic, which killed 14,000 people, was made worse by the cleaning of the sewers by flooding them, thereby pushing the disease into the river. It had been thought that cholera was an airborne microbe and that flushing the sewers would disperse the disease-causing miasmas. The doctrine that cholera was caused by polluted air was one that Florence Nightingale, whose museum stands on the south side of Westminster Bridge, believed until her death.

By the mid-nineteenth century more efficient sewers and the countless flushes of the newly popular water closet meant that virtually the whole of London's two-million-and-rising population was crapping in the Thames even before Thomas Crapper's boast of 'a certain flush with every pull' for his WC later in the century. In 1858, a year after Faraday's letter to *The Times*, there was a long, hot summer and the river began to smell in a way that was almost unbearable for the politicians who had to work next to it. The curtains at Parliament were soaked in chloride of lime, and tons of chalk, lime and carbolic acid were emptied into the river in an effort to control the awful stench, but all to no effect. The 'Great Stink', as it became known, forced Parliament to act. A concerted effort to contain the city's human waste by constructing massive sewers followed, under the supervision of Joseph Bazalgette.

The plans of Bazalgette were in one sense simply to dump London's sewage downstream by intercepting the existing drains, which ran from the northern and southern outskirts into the Thames, with huge west–east running pipes. Three pipes from the north would converge at Abbey Mills before pumping sewage on to Beckton; two pipes from the south, along with a smaller feeder pipe from Bermondsey, would meet at Deptford before pumping sewage on to Crossness. In all 132 kilometres of tunnel was run through built-up areas, and both outfalls were working in 1864, though the scheme was not fully operational until 1875. At first the sewage was put back into the river, but from 1887 to 1998 it was treated before being recycled or dumped out to sea. Since 1998 it has been incinerated.

The sewers have served London ever since, though the system cannot cope during heavy rains and releases twelve million cubic metres of sewage annually into the river. To deal with this storm overflow a scheme is being considered for a relief tunnel, nine metres across and thirty-five kilometres long, through London, under the riverbed. Most of the time, though, thanks to splendid Victorian engineering, visitors to the riverside today can enjoy their time up on the London Eye and stroll along the embankments without the need for nosegays.

CROSS RIVER TRAFFIC

Along with the sewage, London local government is now safely downriver, but the former County Hall on the south side of Westminster Bridge is still able to stir up debate. It was completed in 1933, with additions made until 1974, and it contains a war memorial to former employees to which the property developers, Shirayama Shokusan, who bought the site in 1993 have been less than keen to allow access. The developers also got a very good deal at the taxpayer's expense from the Conservative government in the latter's hasty desire to get shot of the old GLC headquarters.

An odd choice of sideshows is now contained in the building, including an aquarium for a city that already has two (if you count the small one in Kew Gardens), and a museum dedicated to an overrated Spanish artist who never visited the place and that doesn't even contain original works. There was a short-lived football museum that didn't bother to broadcast the results at 4.45 on a Saturday afternoon and, on top of that, there is an amusement arcade and fast-food outlets better suited to scabby promenades in festering resorts than to one of the epicentres of global tourism. Mr Saatchi's gallery, though not the best use of the space, does at least celebrate great modern British art. The key objection to County Hall is that, unlike almost every other official attraction on the South Bank, there is nothing uniquely London or even British about most of the diversions there and it's not as if there isn't an obvious use for the space. This could be a museum celebrating the great people who built the embankments, sewers, tunnels and bridges.

It seems peculiar that building the sewers, having a London-wide system of local government and even constructing bridges were once contentious political acts, but established interest groups fought against each. *The Times*, for example, preferred 'to take its chance with cholera rather than be bullied into health' because it feared the growth of a London-wide government and higher taxes. The City Corporation thought (correctly) that building more bridges would dilute its power, and the erection of the first Westminster Bridge certainly represented the beginning of a shift of focus westwards for London. Today the bridge itself could be described as non-aligned, but it still acts as a conduit for the passionately political to carry their message over the water.

BLACKFRIARS BRIDGE

TO WATCH THE TIDE COME IN

Blackfriars Bridge

At one time this was the point at which the Thames water became officially fresh (rather than saline), so perhaps Blackfriars Bridge is the best place to watch the tide. One can never predict what might be washed up or revealed by the incoming or outgoing waters. In 1962 it was a flat-bottomed Roman barge, and twenty years later, something altogether more sinister, but also Italian.

The first Blackfriars Bridge (originally named the William Pitt Bridge, after the Prime Minister) was built between 1760 and 1769 by a young Scotsman, Robert Mylne, who won a competition to design it under slightly controversial circumstances. Initially the judges chose an entry by John Gywnn, but they later changed their minds because they thought Mylne's design more elegant. Mylne was regarded as a bit of an upstart by the architectural establishment and pretty soon Dr Johnson (a friend of Gywnn's) was heard huffing and puffing about Scotsmen building good bridges being as rare as a dog that juggles.

One of the reasons cited in support of the bridge's construction had been that by opening up the area to the south, the bridge would cause the slums around the north shore to be cleared. It was hoped that improving trade links with what was then the west of the City of London would raise the quality of the neighbourhood and lead to a general increase in prosperity. The area certainly needed gentrifying: Bridewell Prison once stood on the site of the present Unilever building. In its original incarnation it was a palace, then an orphanage, but after the old building was destroyed in 1666 the new Bridewell was a jail pure and simple. On the other side of the road, north of what is now Blackfriars Station, was the notorious Fleet Prison, famous for its debauchery and the cruelty of its warders. This was destroyed in the 1780 Gordon riots and rebuilt shortly after, only to close forever in 1842. Further up the Fleet valley was Newgate Prison, described by Henry Fielding as 'a prototype for hell', which lasted in various guises until 1902. It is now the Central Criminal Court at Old Bailey.

The prisons were located here partly because of the lawyers nearby, based at the Inns of Court, partly because this was the edge of the City of London (Newgate is, after all, a gate to the city), and partly because the district was a criminal ghetto. The notorious Alsatia sanctuary, one of the last outposts of the ecclesiastical law which allowed criminals asylum from pursuing police, stood on the west bank of the river Fleet. It was abolished in 1697, but well into the following century the quarter was a no-go zone for the forces of law and order. It was only just before work on the bridge started, for example, that the rather shady Fleet Marriages were outlawed. These were unlicensed, often bigamous, unions frequently performed by bankrupt priests locked up in one of the prisons – some might say the equivalent of getting married in Las Vegas.

The word 'Fleet' is derived from an Anglo-Saxon word meaning tidal inlet and the area along the Fleet river valley was a slum that no respectable person wanted to live in, as the smelly river carried sewage, dead animals and industrial waste from all points north. It was the largest of London's lost rivers which disappeared in stages from 1732, when the stretch from Holborn Bridge to Fleet Street was covered. This was followed in 1863 by the Farringdon Road section, and in 1765 the lowest reach from Fleet Street to the Thames was topped. The upriver part disappeared when Hampstead started to be developed as a London suburb in the 1770s, vanishing finally through the establishment of Camden Town in 1812. Now the Fleet is barely one metre wide as it whispers apologetically into the Thames underneath the first arch on the north side of the Blackfriars Bridge through a tiny drainage channel.

Before work started on the first bridge in 1760 an elaborate ceremony was conducted involving the burying of money and a plate under the central pile, and Mylne dropped a personal medal into the foundations. There were delays in the middle of the decade after the central arch was completed, because extra funding was required to build the northern approach and trouble in Europe as well as issues in the American colonies meant that money was short. In the interim, timber was put across the approaches to enable people to cross the river, but it was hardly the spectacular portal to the west of the city originally envisaged. The bridge was completed in the end due to an outbreak of disease – in particular a fever that killed several important citizens that was traced to Newgate Prison. When money was released to build a new jail, funds were also set aside for a better embankment and the completion of the bridge.

In 1769, nearly a decade after work started, the bridge with its nine elliptical arches of Portland stone was opened. Despite the delay, Mylne had the cheek to claim that the 303-metre-long (and thirteen-metre wide) bridge cost less than his estimate. It was a beautiful crossing, resting on slender, pointed cutwaters and supported by double Ionic columns. It had an 8.5-metre carriageway with a couple of footpaths either side, and sheltered spots above each pier. Naming the bridge after William Pitt didn't seem like such a good idea any more, as his star had rather waned, and so it was re-christened Blackfriars, after the Dominican monastery that had stood on the north bank. A toll was charged to help cover the costs of construction but the tollgates were destroyed and robbed during the Gordon riots, and charging was abandoned altogether on weekdays in 1785. Charges were kept up on Sundays only until 1811 to help pay compensation to the watermen.

The polluted waters of the Thames and the contaminated outpouring of the Fleet eroded the Portland stone and resulted in the foundations of the bridge being undermined even before the removal of Old London Bridge in 1832 made matters worse. Work began to bolster the piers but by 1840 it was decided to replace the crossing. It was another twenty years before a temporary crossing was erected, pending a more permanent replacement. The old bridge material was recycled and bits turned up all over the greater London area, including a single Ionic column ten metres high and topped with a vase – a monument to Mylne – which ended up in Amwell, Hertfordshire.

The replacement Blackfriars was to have been a three-arch bridge designed by Thomas Page, but the London, Chatham and Dover Railway wanted a railway bridge built nearby. Since the railway bridge required five arches, the road bridge had to have five as well in order to standardize the river flow underneath them. Joseph Cubitt was eventually appointed to design both bridges, and to overcome the effects of the tide, he sank massive iron caissons into the river clay and half-filled them with concrete. He built up his piers on top of these using granite-faced brickwork. The spans are supported by wrought-iron ribs, and the arches themselves are made of iron. Cubitt took no chance with the tide and the bridge itself is more unified, complete and solid than its predecessor, with its features fitting together neatly. The stone piers, with their fat, squat columns, and the linking in of the steel arches with lattice work embellished with a flower motif, bring an overall aesthetic balance to the structure.

The bridge is 281 metres long and, after being widened in 1910 to accommodate trams, 32.4 metres across. It has pulpit-shaped pier heads to suit its name, and on the cutwaters there are columns of polished red granite which are supported by granite blocks decorated by the sculptor J.B. Philip with plants and animals – freshwater creatures and land birds on the upstream side, and marine life and seagulls downstream, reflecting Blackfriars' former position as the tidal turning point. The low cast-iron balustrade completes the Venetian-Gothic effect. There were complaints at the time the bridge was built that it was too ornate, but rarely, if ever, does the decoration on a London structure so accurately depict what goes on around it. Seabirds can be seen today eating large eels just metres downriver of the bridge. The new road bridge was opened by Queen Victoria. She didn't have a great deal of time to admire Mr Cubitt's work as she was jeered vigorously throughout the opening ceremony and had to duck missiles from the London mob in a republican frame of mind, before she dashed up Farringdon Road to open the Holborn Viaduct.

A second railway bridge was built in 1886 (constructed by W. Mills, assisted by John Wolfe Barry and H.M. Brunel) which still takes trains to Blackfriars Station and beyond. The original railway crossing became surplus to requirements by 1923 and, after the superstructure was removed in 1984, now all that can be seen of it are the piers. Shortly after the track was cleared Sam Wanamaker, the man who rebuilt the nearby Globe Theatre, lobbied for an 'art crossing' filled with statues using the abandoned columns of the bridge as plinths. That idea was abandoned but the occupants of Ludgate House, the *Daily Express* building, make good use of the final shoreside support of the railway bridge as outdoor seating. The views downriver from the bridge itself are somewhat obscured by the newer railway bridge, but upriver the Thames sweeps along to Waterloo. It is a famous panorama because many artists painted London from the rooftop of the Albion Mills, built by Matthew Boulton and James Watt in 1786, which formerly occupied the Express building site. It also is very possibly the inspiration for the 'dark Satanic Mills' that William Blake wrote of in 'Jerusalem'. At the northern approach, Unilever House, which curves round from Victoria Embankment to New Bridge Street, is topped by some impressive sculptures. The finest is of a horse and girl by William Reid Dick that casts into even poorer relief the C.B. Birch statue of Queen Victoria somewhat marooned in the traffic below (though at least she's safe from hecklers there). To the left is the former City of London School, with stained-glass depictions of Francis Bacon, John Milton, Sir Thomas More, Isaac Newton and William Shakespeare, among others. On the right-hand side of the north bank the Blackfriars pub has beautiful interiors, ornate fittings, gorgeous mirrors, heavy use of gold paint and pictures of fat monks drinking ale. Underneath the arches of the bridge on the south bank a series of tiled mosaics details the history of bridge building at Blackfriars.

It was below the arches on the north shore in June 1982 that the body of an Italian banker, Roberto Calvi, was found hanging with bricks in his pockets. As a senior figure at the Banco Ambrosiano, Calvi was responsible not only for mob money-laundering but also the Vatican accounts, as well as (if you believe the conspiracy theories) Nazi gold deposits and payments to a suspicious Italian Freemason group called P2, which helped to fund fascist terror groups in the 1970s. He was initially thought by some to have been executed by the Freemasons for betraying their secrets. The combination of Mafia, Freemasons and the church caused a riot of conjecture and there are books, a film and hundreds of websites devoted to the Calvi case. Calvi's knowledge of certain dubious Vatican investments (something for the weekend, sir?) might have meant that the priests who orated over his funeral did so with a certain satisfaction. The initial verdict of suicide was bewildering to many people, not least Calvi's family who, after two decades of campaigning, had the body disinterred. A new report into Calvi's death then put the blame where many suspected it should have gone in the first place. Mafia turncoats have accused the London boss Francesco 'the strangler' Di Carlo of killing Calvi to punish him for absconding with some of Cosa Nostra's cash.

Many refer to Calvi as God's Banker, but he was really the middleman; the person who controlled the Vatican accounts from God's end (as it were) was actually Bishop Paul Marcinkus, the American head of the Vatican bank. Calvi's death at Blackfriars might, at least, have helped him in the next life for it was believed that a person buried in the dark robes of the Dominicans would never lose their soul to the devil. As to whether being wrapped in the black waters of the Thames under a bridge named *after* the Dominicans works the same trick, you'd have to ask Roberto, and his lips are sealed, which is exactly what some people wanted.

In an entirely different murder story, Shakespeare, who lived in Ireland Yard near the site of the future bridge, wrote 'fair is foul, and foul is fair: Hover through the fog and filthy air'. Terrible London fogs are possibly what he had in mind and the description certainly fits. As the south-east's forests shrank in the twelfth century, large deposits of sea-coal from the north-east coast provided the cheapest alternative to wood. Along the Fleet valley, Old Seacoal Lane was where coal was docked, and nearby Newcastle Close signifies where it came from. This coal does not burn efficiently and a good deal of its energy is spent making smoke, not heat. A Londoner coined the term 'smog' in 1905 to describe the city's insidious combination of natural fog and coal smoke, which is also known as a 'pea-souper' or 'London particular' and is the reason some British people still refer to London as 'the (big) smoke'.

By the 1800s, more than a million London residents were burning soft-coal, and winter 'fogs' became more than a nuisance. In 1879–80 a particular lasted from November to March and in 1952 a four-day fog officially killed 4000 Londoners, though a later study claims the actual total was 12,000: the majority of those who died did so from respiratory illnesses, though a significant number lost their lives falling off bridges or into unseen traffic. Things were so bad that the blind helped the sighted through the streets and in 1956 Parliament passed the Clean Air Act to reduce the burning of coal.

The fogs coated everything in the city with grime and many buildings blackened by the soot were only restored to their intended colours in the late twentieth century. Since the air, like the Thames water, is currently purer than at any time since, some claim, the Middle Ages, repair work and redecoration are now worthwhile. The jaunty use of colour on the metalwork of Blackfriars Bridge today is a fine example of the restoration work that has gone on throughout London. Blackfriars now rejoices in a lively red, white and gold, with gold emblems fixed into the supports. For a variety of reasons (including wartime camouflage), many of London's bridges were painted a dull grey in the early twentieth century.

The bridge and London's air have been cleaned up, but what about the area around Blackfriars, which was one of the original reasons for the crossing? Looking at the neighbourhood today the purification seems to have worked, even if it took some time. A clutch of terrible prisons, some slums, a smelly river and a criminal nest were all cleared long before the journalists, who arrived in the sixteenth century, began to flee the Fleet Street area in the 1980s, led by Rupert Murdoch. Some might say that just leaves the lawyers to depart before the area can be declared completely free of corruption.

LAVENDER HILL MOB

Battersea Bridge

During the erection of the first bridge at Battersea, which opened in 1771, a large quantity of human bones and weapons, dating from the Roman occupation, was found on the riverbed. There's a popular notion that this is where Julius Caesar crossed the Thames to scrap with the Catuvellauni tribe on the north bank. It was the obvious choice for the first bridge to link Chelsea on the north bank with Battersea to the south: roads already led to the spot as a ferry had operated for centuries from nearby, and before that it had been a fording point. Initially the crossing benefited the community on the north bank by helping Chelsea to evolve into a coherent settlement, rather than a disparate collection of dwellings. For nearly a hundred years the Chelsea Bridge, as it was once known, was the only crossing in an area that today has three bridges within a couple of kilometres of each other.

John, Earl Spencer, who promoted and sponsored the bridge, thought that it would be safer than the old ferry and would make money for anyone who invested in it. Not least the Earl himself who hoped it would develop the area to the south-west where he owned quite a bit of land. By 1766, fifteen people had put up £1500 each to help the Earl build the bridge, which sum was not enough for the elegant stone crossing originally planned. It did stretch to building an 8.5-metre-wide wooden bridge designed by Henry Holland and completed by Mr Phillips, carpenter to George III. In 1772 the first wagons were allowed across the bridge after a year of pedestrian use only. In 1795 iron girders were inserted to strengthen the structure and in 1799 oil lamps were added to make it safer to cross at night. One-metre-high iron railings replaced the old wooden fence in 1824 but none of these measures made it any more attractive and it is described in 1865 in *Cruchley's London: A Handbook for Strangers* as 'a deformed, dangerous, and hideous-looking structure of wood'. *Punch* magazine was even more scathing twenty years earlier, saying of the bridge that the:

> curious old pile is a singular expression of the barbarism of Gothic ages. It was built for the purpose of enabling persons to go over it, and little regard has consequently been paid to convenience of those who desire to go under it. It has the appearance from the river of an old fortress, and is garrisoned by a single toll-keeper.

The Earl and other investors were, at least, able to recoup some of their stake through the tolls, which cost one shilling for a carriage, and a halfpenny for pedestrians. In 1858 competition arrived with the opening of what is now known as the Chelsea (formerly the Victoria) Bridge just downriver. This caused a severe drop in revenue, which was compounded by public concerns about the safety of the Earl's bridge. It was bought under compulsory purchase by the Albert Bridge Company in 1874 and five years later it was requisitioned by the Metropolitan Board of Works. Tolls were immediately removed, in 1883 vehicular use was discontinued, and within four years the bridge was demolished.

It wasn't just the safety or strength of the bridge for road traffic that led to its destruction. Henry Holland's original design featured eighteen closely positioned wooden piers, which caused problems on the river itself. It was an accident black-spot, and cargo boats regularly collided with the bridge at night or during times of fog or poor visibility. Furthermore the restriction caused to the flow of water by the many arches created a weir and made it difficult and dangerous for boats to pass under, or 'shoot the bridge'. It was also, after a notorious murder in 1844, deemed quite dangerous to cross on foot. Sarah MacFarlane had been butchered on the bridge by Augustus Dalmas, apparently in view of one of the toll collectors, who didn't intervene because both parties had paid their money.

But artists loved the old bridge. It was, along with the Battersea Reach of the Thames, a favourite subject for James Whistler and Joseph Turner, both Chelsea residents, to paint. Turner depicted the Thames in the classical, romanticized tradition more associated with beautiful Italian rivers than the working London one. Whistler was even bolder in his take on the river and celebrated the beauty contained in the polluted urban landscape. Whistler's paintings of the bridge (and his entire 'nocturnes' series) had a huge influence on modern painting. These seemingly simple pieces imbue the subject matter with a remarkable combination of romance and melancholy that opened up a new world of expression in painting and heralded the beginning of abstract art. In 1899 French Impressionist Claude Monet began a series of paintings of the Thames influenced by the work of Whistler and Turner, though his most famous London paintings are of Charing Cross, and Westminster and Waterloo bridges. A rather less obvious link between Turner and Whistler is provided by another British painter, Walter Graves, whose best-known work is of Hammersmith Bridge. Graves, as well as being a painter, was Whistler's boatman, and his father had held the same position for Turner.

Joseph Bazalgette, however, though no philistine, had no time for artistic shilly-shallying where bridges were concerned. His engineers found that the Earl's bridge was unsafe and unable to cope with its current traffic levels. In view of the likelihood of increased and heavier traffic due to London's development and the removal of the tolls, he decided to demolish the bridge. At first the Earl's bridge was not going to be replaced at all, then it was thought better to keep the old crossing in place as a new one was constructed alongside it. Finally the re-alignment of the road system approaching the Thames meant that any new crossing would have to cut at a diagonal through the old.

Bazalgette completed the stone-based replacement by 1890, and the bridge was re-christened Battersea. This name derives from Badric's Ey – Beaduric being the chieftain who once controlled the area, and 'Ey' meaning island, suggesting an area surrounded by marsh or water. The first mention of the name Battersea (as Batrices Ege) appears in a charter of AD 693. In later centuries Battersea became a market garden for London, after successive draining of the marshland produced fine farmland famous for (among other things) lavender, hence Lavender Hill further inland. It was also well known for asparagus, which was sold in Battersea bundles, which bear no relation to Camberwell carrots. The parish church of St Mary is near the bridge, and was the focus of the original village. In its present incarnation the church dates from 1775: William Blake was married there and Benedict Arnold, best known for helping the British in the US War of Independence, is buried in the churchyard.

Described by one present-day architect as 'Joy Division to Albert Bridge's Kylie', Bazalgette's serious, solid bridge is supported by four granite piers on concrete foundations. Five cast-iron arches ribbed with wrought-iron cross-members, for extra support, form each span. It is currently painted green and inset with small gold adornments, like a Bendicks chocolate box. The lattice work on the balustrade is very fine and almost Moorish in effect. There is an array of lights along the top, with a trident supporting three lamps followed by two individual light spurs at regular intervals. It's quite a low-slung bridge, with a cool dignity and comfortable strength, befitting the genteel areas it links and the many notable citizens who once lived in the neighbourhood.

The lighting on the bridge is attractive but has failed to rid it entirely of the problems that bedeviled its predecessor. There have been a series of closures after boats collided with the bridge and in 1979 there was another murder. Still, the bridge is a safer way of crossing the Thames than either the old ferry or bridge – or indeed the method chosen by Madame Genevieve Young, 'The Female Blondini', who crossed the river on a tightrope near this point from Cremorne Gardens, where the Lots Road Power Station building is today. These gardens opened in 1840 and boasted a theatre, banqueting hall and grottos. Other attractions included balloon ascents, with a leopard on at least one occasion, and circus sideshows. The local Baptists led a campaign against the gardens, which they saw as a nursery for vice.

As a result the owner John Baum lost his licence in 1871; even after regaining it some years later he was left bankrupt and the gardens closed in 1877.

The oldest annual contest in continued existence anywhere in the UK, the race for Doggett's Coat and Badge, still takes place near the bridge. Thomas Doggett was an actor and theatre manager who lived in Chelsea. In his will, he remembered the watermen who had rowed him home in all weathers, bequeathing a sum of money to be used for an annual boat race up the Thames from London Bridge to Chelsea (Battersea) Bridge. The competition is administrated by the Worshipful Company of Fishmongers and the first race was in 1721. Traditionally the participants were drawn from the ranks of Thames watermen and lightermen and the prize is a scarlet coat with a silver arm badge that the winner gets to keep for the year.

CROSS RIVER TRAFFIC

For centuries boats were the preferred means of transport for the quality to the City and Whitehall, though the last stop of the river boat today is at Cadogan pier nearer Albert Bridge. The journey east cannot even be taken by rail as the nearest railway line runs the wrong way, across London from north to south. When Battersea Rail Bridge opened in 1863, just upriver of the road crossing, the track was supposed to link the southern systems running from Waterloo, Clapham and Victoria with the northern ones operating out of Paddington and Euston. The bridge itself has four brick piers, faced with Brambly Fall stone set on concrete foundations and, since the 1990s, 3500 cubic metres of lightweight foamed concrete. This was pumped into the bridge supports and allowed bridge strengthening to take place with practically no disruption to the trains. The foamed concrete was pumped in one-metre layers and fully encapsulated the decaying steel girders supporting the bridge deck. The bridge itself has five elegant spans and attractive lattice work linking the wrought-iron girders. The rail line is heavily used by freight but not so much by people, which is a shame, as speed restrictions on the rail bridge mean that there is plenty of time to admire the river.

The Counters (or Chelsea) creek flows into the Thames between the road and railway bridges. In 1829 it marked the start of the Kensington canal that ran for three kilometres of the creek's length. Prior to that a fording point gave its name to Stamford Bridge (Sandy ford), the home of Chelsea Football Club, and the seven hectares from the creek to the old railyard has become the Chelsea harbour complex. This development and others on the north bank, such as the former Lots Road Power Station, are being matched on the south side by the Albion riverside development. The Albion stands between Albert and Battersea bridges and has been described, admittedly by estate agents, as one of the most desirable places to live in London, thanks to its design and location.

LAVENDER HILL MOB

Battersea Bridge

The Albion is rather impressive and makes good use of the area without blocking off the river. There are areas of early gentrification (notably in Wapping) where the right of way is restricted, but most of the more recent residential wharfside developments have included bars, restaurants and other amenities that attract people to the area. Developers have even laid down cycle paths and constructed walkways under the bridges, aids to public access and enjoyment that several councils could do well to mimic. Even so some politicians are concerned that London is returning to a pre-Victorian model, when the houses of the wealthy lined the Thames to the west, and that this will result in polarized communities of rich and poor.

The Albion and other developments are part of a massive expansion of the population along the Thames from Limehouse to Wandsworth since the late 1980s.

Some of these are on former industrial or underused land, others are conversions of warehouse properties. It could be argued that the bridges initially encouraged the development of the swag-bellied suburbia around London, and this was certainly one of the intentions of Earl Spencer, patron of the first Battersea Bridge. Many of the new settlements however, particularly in the west, are clustered around the bridges themselves. And, like the health clubs that are ubiquitous features of these developments, the riverside housing stock is now providing a belt-tightening exercise for the Greater London area by pulling the city's population closer in to the river.

GAZING ON PARADISE

Waterloo Bridge

London Bridge may be the most famous bridge in the world and Tower the most recognizable. There may be poems about Westminster, and the Millennium can wobble all it likes, but Waterloo is London's most evocative bridge. This is the crossing of romantics and suicides, the bridge built by women, the one with the best views that opens up the whole potential of London. It doesn't really matter that the 1931 movie *Waterloo Bridge* was filmed on a back lot in California, because Waterloo Bridge is a state of mind as well as a river crossing. As Ray Davies captures so brilliantly in his song 'Waterloo Sunset', the bridge is a place out of time, neither land nor water. It is a crossing point not just to another side, but to a different world of fresh possibilities, from the (then) poverty of south London to the bright lights of the West End, and the opportunity of a new life as the sun sets on the old.

The emotionally charged 1940 remake of *Waterloo Bridge* was shot on location and stars Vivien Leigh as a young ballerina who falls for Robert Taylor, a Scottish officer. They meet on the bridge during an air raid in the First World War (an attack by the menacing Gotha aeroplanes towards the end of the war, rather than the ponderous Zeppelins of earlier raids). Leigh loses her position at the ballet school after spending a last evening with Taylor before he heads to the front, and when he is reported missing presumed dead, her increasing poverty drives her to prostitution. She's plying her trade at Waterloo Station when she spots Taylor. He thinks she's come for him and she can't bear to tell him the truth. Plans for the wedding are made but she cannot live with herself and eventually she throws herself off the bridge. The film starts and finishes on Waterloo Bridge, allowing the viewer to see a mock-up of the original bridge and the temporary structure of the replacement, then under construction. It opens with Taylor pausing to reminisce during the Second World War, and the final scene shows him tossing a lucky charm Leigh had given him into the waters. The film has everything: love, life, opportunity, despair and hope, which pretty much sums up the bridge.

On the northern shore, the River Terrace that links directly to the bridge leads to Somerset House, which formerly held the records of all the UK's births, deaths and marriages, tying the bridge to the lives of almost every man, woman and child in the country. Next to Somerset House, on the bridge's carriageway, is the Inland Revenue building, cementing the bridge to the second of Ben Franklin's great certainties of life. Heading north, the carriageway passes above the old Kingsway tram depot and sweeps across the Strand onto Aldwych, or straight ahead to the theatres and bars of Covent Garden and the West End. To the south, the iconic Imax Cinema rises out of the old bull ring. On its left is the Church of St John the Evangelist and King's College and, further on, the thriving area around Waterloo Station. The station was built in 1848 and reconstructed by 1922 to include the beautiful victory arch. The Eurostar terminal for journeys to Paris and beyond was added in the 1990s. The bridge leads to fantasy, foreign travel, education and redemption, making a trip across it in either direction crackle with possibility.

The first Waterloo, or Strand Bridge as it was formerly known, was opened on 18 June 1817 by the Prince of Wales. As this was the second anniversary of the Battle of Waterloo, the Duke of Wellington was present for the official ceremony. The bridge had been completed by John Rennie after six years' work, and it was universally admired for its Grecian Doric columns, cornice balustrade and lofty beauty. Baron M. Dupin described it as 'a colossal monument worthy of Sesostris and the Caesars'. *Cruchley's London in 1865: A Handbook for Strangers* said that its 'simple grandeur cannot fail to excite the stranger's admiration'. To Canova it was simply 'the noblest bridge in the world'. It was unique at the time for having a level roadway supported by nine semi-elliptical arches with a span of equal size (forty-two metres) and elevation. Built mostly of Cornish granite, except the balustrades, which were Aberdeen granite, the bridge was 378 metres long over the river and thirteen metres wide. The total length including the approach to the Strand and causeway on the Surrey side was 749 metres. The extra roadway was necessary in order to achieve a level surface on the bridge and raised the cost considerably, as the Strand Bridge Company had to acquire many adjoining properties.

The bridge towered over the river and was a huge testimony to London's wealth and power. Its central location meant investors expected a healthy return from tolls despite the high construction costs. Unfortunately the tidal scour that occurred after Old London Bridge was removed (ironically by Rennie's son) damaged the foundations. Every pier had to be reinforced between 1882 and 1884 after the bridge had been bought out by the Board of Works (who paid a record fee for it). By 1923 the three central piers were sinking and attempts were made to pump concrete under them. When this didn't work the bridge was closed.

There followed a ten-year battle between those who wanted the bridge restored (the Royal Commission, the government, the Royal Academy, and virtually every heritage and art society) and those who thought a new bridge should be built (the London County Council). Even if the bridge were restored, the LCC argued, it would not be able to cope with modern traffic demands, and the Council decided to press ahead with a new bridge despite the fact that the government initially withheld any grant money towards it.

Sir Giles Gilbert Scott was commissioned to design the new bridge and work began in 1937, although the official foundation stone, cut out of a stone from the old bridge, was not laid until 1939. Coins, postage stamps and all the newspapers of the day were placed within a copper cylinder and lodged in a cavity under the stone. Delays occurred because of the Second World War even though the bridge was deemed a priority and, with few men available, women carried out much of the construction work.

Plans to sell off bits of the old bridge didn't fully materialize, though balustrade struts were sold at £1 each and at least one of these ended up in a garden in Dorking, where it supports a sundial. Rather bizarrely, wooden pilings from Rennie's crossing were so well preserved that they were pulled out of the clay and used to construct new coaches for the London, Midland and Scottish Railway Company.

One section of balustrade and two of Rennie's Doric columns are preserved in the southern abutment as well.

Despite being damaged by German bombers on several occasions, the 'Ladies' Bridge' was opened to pedestrians and two lanes of traffic in 1942, although it was not fully completed until 1945. In view of the arguments with the architectural establishment and the government over its rebuilding, Waterloo is very much the people's bridge and it is appropriate that it was opened by a member of the public, Charlie Barnard, a steel fixer from Ealing. He removed seven red flags at the Surrey end and the bridge was declared open, prompting a race to be the first across that was won by Leonard Mitchell, a sixteen-year-old schoolboy from Balham.

Waterloo Bridge is a cantilevered five-span crossing and the first to be made with reinforced concrete beams, though it is faced with Portland stone. It is the longest bridge in London and a triumph of confident simplicity and harmony as well as being the first to incorporate electric lights, attractive though the old gas ones with a cannon motif had been. The great sweeping arches appear more majestic because of the flatness of the bridge itself and manage to give the solid structure the illusion of lightness. Its austere beauty and lack of decorative features help to emphasize the amazing sights up and down the river from the bridge. Being able to see through the balustrades adds to this grandeur: vehicles and people appear to glide over it. Even the underside has more going on than any other crossing, with the National Film Theatre (and its café) firmly rooted on the southern shore along with a second-hand book market. Under the bridge there is a fantastic symmetry to the concrete as it arches away to the centre, giving a sense of order and balance – in direct contrast to the nearby skateboarders who skid chaotically along the South Bank.

Downriver on the south side are Gabriel's Wharf and the Oxo Building. Formerly a Post Office power station, the latter was bought by the beef stock cube company which, in the 1920s, built the tower with its famous stained-glass windows that spell out the company's name (as a means of getting around restrictions on neon advertising). The adjoining site was derelict until 1977 when the Coin Street Action Group successfully fought off developers' plans and erected affordable housing, small businesses and public spaces. Upriver on the south side is the South Bank Arts Centre, made up of the National Theatre, National Film Theatre, Royal Festival Hall and the Hayward Gallery. To the north and downstream (beyond Somerset House and the Savoy) are the Inner and Middle Temples, the centre of British law since the fourteenth century. Beyond that there is Blackfriars Bridge and St Paul's and a view off over the city towards Canary Wharf. It is hard to know whether it is these views or the bridge itself that makes it a popular spot to propose marriage.

There is also a darker side to the bridge, which has witnessed suicide, ghosts and political assassination. Georgi Markov was a Bulgarian dissident who escaped his country and worked for the BBC World Service at Aldwych. On 7 September 1978 Markov was crossing Waterloo Bridge on his way to work when a chap in a bowler hat caught his leg with an umbrella. Markov got to the end of the bridge, collapsed and died a few days later. The tip of the umbrella had contained the poison ricin and during the autopsy doctors found a small platinum pellet embedded in his calf. In an astonishingly Ian Fleming way the KGB had got their man.

Markov is not the only person to have met his end on or near the bridge. It seems to have been the suicides' bridge of choice (although Blackfriars was a stiff competitor) – at one point in the nineteenth century these averaged forty a year. In 1860 Charles Dickens wrote of the bridge in his weekly periodical *All Year Round*:

> But the river had an awful look, the buildings on the banks were muffled in black shrouds, and the reflected limbos seemed to originate deep in the water, as if the spectres of suicides were holding them to show where they went down. The wild moon and clouds were as restless as an evil conscience in a tumbled bed, and the very shadow of the immensity of London seemed to lie oppressively upon the river.

Thomas Hood composed the poem 'Bridge of Sighs' about a suicide from Waterloo Bridge, although the case that inspired it was actually that of Mary Furley, who had drowned herself in the Regent's Canal. More suicides occurred in canals and parks than from bridges over the Victorian era as a whole, but the drama of a bridge jumper seemed to capture the public imagination better. Whatever the reality of the suicide demographic (and by 1840, fifteen per cent of London's suicides *did* occur on Waterloo Bridge), a lasting link was made between suicide and the bridge by the repeated portrayal of the subject by novelists, artists and even historians and travel writers. Not all artists represented the bridge in a bleak light: some, such as Claude Monet in his *Sun in the Fog* painting, showed it instead in a hazy one. For most though it was too good an image to waste (a beautiful woman, bridge arch, moon, river, historic buildings, a church in the background to give a moral dimension) and it became an easy shorthand that everyone could understand. Several successful plays were based around it including Charles Selby's melodrama *London by Night* (1844) and sometimes, in a bid for extra drama, a suicide would be re-enacted onstage. This was not always a good idea, as actress R. Honner found when she missed the mattress on a four-metre leap from a 'bridge' during the play's December 1843 opening, and was very seriously injured.

American diver Sam Smith was less fortunate and died at Waterloo Bridge in 1840 after a lifetime spent successfully plunging off tall buildings, docks and cliffs all over the world. More gruesome was the discovery of severed human remains near the bridge in a carpet bag in the nineteenth century. The bag was lowered from balustrade to abutment and deposited there by a woman or, as a witness put it circumspectly, 'a person in female attire'. Human remains were also found in the mid-twentieth century and this latter discovery was followed by a ghostly apparition appearing near the bridge for weeks afterwards. On the embankment at Cleopatra's Needle near the bridge, a dark figure – apparently the ghost of a suicide – is still often seen leaping over the wall near the monument. There is never a subsequent splash, but there are reports of phantom laughter coming from the area. It would appear that not all the suicides are resting easy.

These grim tales shouldn't detract from the most important thing about Waterloo Bridge, which is that this is the place where Terence Stamp and Julie Christie cross the Thames in 'Waterloo Sunset'. These 1960s icons symbolize the sense of wonder at Swinging London, a beautiful city being reborn. That the greatest love song to any city anywhere should be associated with a view from a bridge is perhaps fitting. Because the Thames brought London into being and the bridges caused its expansion, and there is nothing to be frightened of as long as we can stand on Waterloo Bridge gazing at paradise.

HUE AND CRY

Southwark Bridge

Southwark, pronounced 'Sutherk' (no enunciated W's round here my son!), is one of the oldest parts of London, though its other name, 'the ward without the bridge', suggests a place slightly set apart. In Saxon times the area was known as Suthriganaweorc, or fort of the men of Surrey, and the name derives from the old south work or southern defensive fixture. Old Southwark was a fairly lawless place, and for centuries people travelled across the water in order to indulge in pastimes more strictly policed on the north bank. Many criminals retreated to Southwark after pulling jobs in the city, so perhaps it should not be a surprise that Southwark

Bridge's best-known cinematic roles have been in crime-related films, starting with the lovely *Hue and Cry*.

Southwark Bridge's simple yet elegant Georgian forebear, designed by John Rennie, was a footbridge on granite stone piers and abutments with three magnificent cast-iron arches. It linked Cheapside and Mansion House to the north with the industrial Bankside. A contemporary commentator described it as 'charming, graceful and fairylike, spanning the waters in three moves combining engineering skill with architectural elegance', and it is also the iron bridge cited in Charles Dickens's *Little Dorrit*. Work on it started in 1814, and for its opening at midnight on 24 March 1819 the bridge was brilliantly illuminated, temporarily blinding the public to the fact that its cost had risen by more than a quarter during construction. The Southwark Bridge Company was in a mess financially before the crossing was even completed and Rennie had to sue for his fee. The Rotherham foundry, Joshua Walker, that supplied the iron for the 203-metre-long crossing was bankrupted.

Southwark Bridge Company had been convinced that there was money to be made despite opposition from the City and the Thames Conservatory Board, who felt that an additional crossing was unnecessary. Pressure from Thames conservation groups meant that the bridge had to leave 200 metres of clear water so as not to severely disrupt the flow of the river and to allow the safe passage of ships. River traffic may have flowed under the bridge but it was too narrow to accommodate non-pedestrian traffic over it. This, along with poor service roads and its arched shape, meant that it never really made enough money for the investors, despite the heavily populated areas on both sides. Most people still preferred the toll-free and flatter bridges – London and Blackfriars – on either side, making them incredibly congested. In 1866 the Bridge House Estates bought Southwark Bridge after leasing it for two years and removed the toll charges which dramatically increased traffic, but Southwark still remained relatively underused, particularly after walkways alongside Cannon Street Railway Bridge provided extra competition when they were opened in the late 1860s. These walkways were closed when the track was widened, and subsequent plans to reopen them as a covered footbridge alongside the bridge in 2001 were abandoned.

The Bridge House Estates decided fairly swiftly to build better approach roads and a new, broader Southwark Bridge. They were less rapid in acting on the decision: demolition work did not begin until 1913 and, with the First World War disrupting the work and the supply of materials, the current bridge was not completed until 1921. The original cast-iron structure was rebuilt in steel to a design by Mott and Hay as engineers and Sir Ernest George as architect. Its five arches and four turreted piers are aligned with those of Blackfriars Bridge in order to minimize the effect of the tide and reduce cross-currents. George designed the unusual turrets, though they appear to be based on sketches of a proposed Southwark Bridge of 1877. These sketches, by Horace Jones, show statues in what look like sentry boxes that rest on the piers and also on squarer parapets above them. The actual bridge has these 'boxes' above the piers but they remain empty, and there is a circle on top of the turrets that serve as a viewing platform. The distinctive, but truncated, lamp standards mounted on the balustrades are spaced twenty metres apart. The lamp columns originally had three lanterns but unfortunately these dazzled passing boats, and in 1956 they were cut off.

None of the features – the long sentry boxes on the piers, the mounted lights or the circles in the turrets – are unsightly in themselves, but taken together they make for a slightly schizophrenic crossing. There are echoes of the never-built

Victorian bridge, yet the design itself is Edwardian and the work was only completed after the First World War – so in socio-architectural terms the bridge could be an interesting reflection of the confused state of Britain after the war. But the fusion of Arts and Crafts with the baroque makes it a bit of a mess architecturally. The long sentry boxes topped by low viewing points and combined with taller lights detract from the sweep of the arches and make them look unbalanced and slightly awkward. There is nothing wrong with the bridge's rather attractive gold and pale green paintwork though.

London bridges in general may have helped to end the divisions between different areas of London by joining them together, but Southwark Bridge still offers clues as to what those divisions might have been. On the north bank are Thames House and Vintner's Place, on which is a fantastic relief of a topless lady with bunches of grapes covering her modesty and two sheep in attendance at her feet. Underneath the bridge on the north bank Thames path are tiled representations of the building of the old bridge and plans for the proposed replacement. These all, apart from the naked lady, suggest industry, progress and planning.

Underneath the southern walkway, on the other hand, is a chiselled depiction of one of the frost fairs which occurred when the Thames froze solidly enough to hold people, stalls and, on one occasion, a bull ring. The etchings show the good people of London eating and drinking and carrying on a huge party on the river itself. The wider river, severe weather and the effect of Old London Bridge, which dammed up freshwater behind it, made these events possible. Fairs are recorded in 1564–65, 1683–84, 1715–16, 1739, and 1813 into 1814. The Thames in London has only frozen once since then, in 1895, when it was strong enough to support people, but not a carnival. By the time of the last frost fair Southwark's attractions had been mostly closed down but for centuries it had been the place where people went for a bit of fun and entertainment. In 1587 speculators built a new playhouse, the Rose, at Bankside, which was followed by the Swan (1595), the Globe (1599) and the Hope (1614), designed to show animal-baiting and plays. Although relations between some locals and the thespians were strained on occasion, it was not until 1642 that Puritans ordered the suppression of the playhouses. Oddly, one of Southwark's other attractions, bear-baiting, continued until 1685 and was not declared illegal until 1835.

Bear-baiting was promoted as a healthier alternative to the area's other great lure, one that largely disappeared with the closure of the theatres. Some sources argue that Southwark's brothels date back to Roman times, others to the Middle Ages; certainly by the end of the fifteenth century there were eighteen licensed brothels on Bankside. In 1546 they were suppressed on the orders of Henry VIII, though many, including the Cardinal's Cap, had reopened by the reign of Elizabeth I, and the Barge, the Bell, the Cock and the Unicorn all survived into the seventeenth century. The Bankside stews (as they were known) offered a range of services, and ladies. One promised:

> beauties of all complexions from the coal black cling fast to the golden locked insatiate, from the sleepy eyed slug to the lewd fricatrix

There was also quite a variety of establishments, from the magnificent and costly Holland's Leaguer in Paris Garden, to private houses, alongside the many ladies who worked the streets and alleys. In 1650 Parliament passed a law that made adultery a felony punishable by death and fornication a crime punishable by three months' imprisonment, which had some effect on trade, but an idea of the scale of prostitution in Southwark can be gleaned from John Stow's *Survey of London* (1603). He describes the Cross Bones burial site in Southwark as being 'dedicated to single women forbidden the rites of the Church as long as they continued a sinful life'. No other area had its own graveyard for prostitutes.

The toms eventually followed the actors and crowds to the new theatre district of Covent Garden, and by the eighteenth century Bankside had become an industrial area of wharves, warehouses and overcrowded tenements. It is now home to many offices, but it is art and tourism that have brought the visitors back in huge numbers. The rebuilt Globe Theatre was opened a short distance from its original site in 1996 and the Clink Prison Museum, the Vinopolis Museum of Wine, and the revamped Borough Market have all helped. But it was the Tate Modern art gallery in the former Bankside Power Station building that really transformed the waterfront and returned Southwark to being one of London's premier entertainment areas. Tourism is responsible for thirteen per cent of all jobs in London and adds £8 billion annually to the economy and, with a continuous stretch of attractions along the river east of Tower Bridge, Southwark gets a large share of this.

Despite all this activity, Southwark Bridge remains central London's least-used bridge. Road traffic is under a third of that of Blackfriars or of London, and pedestrians tend to approach Bankside along the riverbank or otherwise cross the Millennium Bridge to reach the area from St Paul's. This is a pity, as the bridge offers fine views of the various Bankside attractions, and around its north side are little-used alleyways and sections of Thames path that offer a less crowded stroll along the riverbank. The desolate Walbrook Wharf, where huge barges are loaded with the city's rubbish, even gives a taste of a working dock area, while the nearby Walbrook

Stair offers a reminder of where people once boarded boats to cross the water to old Southwark. This area is the site of the old Saxon port of Queenshithe and the entry to the Thames of the river Walbrook.

Walbrook Stair leads down under Cannon Street Railway Bridge (formerly the Alexandra Bridge), which was completed in 1866, expanded in 1893 and redesigned in the 1930s. In 1981 some of the original ironwork was replaced. It is a very industrial affair set on twenty-four concrete-filled fluted cast-iron Doric columns and painted an olive green. It almost has the appearance of a temporary military fixture until you get close up and see how robust it is, squatting inelegantly over the river, and affording no views. Just upstream of the bridge, on 20 August 1989, the dredger *Bowbelle* collided with the crowded pleasure cruiser *Marchioness*. Police patrol boats were swiftly summoned and arrived on the scene within minutes. They were assisted by the crew of a passenger launch, which was fortunately in the area, and were able to rescue eighty-seven people from the river. A grimmer task followed as officers from the Marine Support Unit (MSU) recovered fifty-one bodies, twenty-four from the wreck and twenty-seven from the river. The Thames Barrier was closed during this operation to slow down the effect of the tide on the river, but it was days before all the victims were recovered, as bodies that land in one part of the Thames can wash up almost anywhere.

Pulling bodies from the river is the bleakest part of the river police officer's work, though large-scale disasters like the *Marchioness* tragedy are very unusual, and suicides from the bridges less frequent in recent years. The number of corpses pulled out of the river each year has declined over the past decade from around 100 to under 40. This is not because Londoners are any less depressed or murderous, but increased mobile phone usage has meant that would-be suicides are reported more rapidly. This means that the police are more likely to catch people alive rather than having to take their bodies to the Wapping headquarters for identification.

The present station for the Marine Support Unit at Wapping was built in 1908 on the site of the original building that opened in July 1798, and it is now the only MSU base on the river. Founded in 1789 the Marine Police Force was the first properly organized and uniformed constabulary in Britain, set up jointly by the government and the West India Dock Company, who paid most of the bills. The Marine Police Force was brought fully into the public domain by the Marine Police Bill of 1800 and merged with the Metropolitan Police (set up in 1829) as the Thames Division in 1839. In 2001 the Thames Division was renamed the Marine Support Unit, whose remit is to patrol London's rivers and vast network of inland waterways.

CROSS RIVER TRAFFIC

The MSU's primary concern, however, is the Thames, and their main focus, aside from general crime and safety, is terrorism. Their other role is as a community police force for river users even if few people cross the river by boat today and fewer still for the guilty pleasures that the south bank once specialized in. The current sobriety and financial stability of twenty-first-century Southwark is symbolized by the *Financial Times* building at the south-east corner of the bridge. Some north Londoners still joke about needing a passport to go south of the river but Southwark is a booming, integral part of central London. There are even plans to change the area's postcode from SE1 to SC1 (for South Central) in recognition of the upgraded status of this part of town. Whether anybody will ever manage to get the silent 'w' reinstated is another matter altogether.

WEAK
BRIDGE
7.5T
mgw
HAMMERSMITH
BRIDGE

SHE'S IN A SITUATION

Hammersmith Bridge

The ballad 'She's in a Situation' is about a married man's affair with a serving girl who lives near Hammersmith Bridge. The song details all the larks they get up to, behind his wife's back. It could be said that the man was following in a royal tradition, as Nell Gwyn and Charles II used to meet at the Dove and stroll the nearby Hammersmith riverbank as well. The pub was also favoured by Hammersmith resident, designer William Morris, and the waterfront at Hammersmith is one of the best bits of the river to visit to have a lazy evening pint, watching the sun go down. Gazing across to Barnes you can almost delude yourself that you are no longer in London: everything looks pleasantly green and the river is alive with rowers.

Both the public and the architectural profession were very excited when it was announced that Hammersmith Bridge was to be the first suspension bridge across London's Thames. To mark the start of work on the bridge in May 1825 there was a Masonic ceremony, with the Duke of Sussex, Augustus Frederick, presiding in front of the Grand Lodge and a large crowd. He performed a ritual that involved the fixing of an engraved brass plate that praised the builders and the designer, and the placing of gold coins and a silver trowel in one of the coffer dams. As this was put in place the Duke completed the ritual by making a further offering and saying:

> I have poured the corn, the oil and the wine, emblems of wealth, plenty and comfort. So may the bridge tend to communicate prosperity and wealth.

An outline of this bridge is in St Paul's Church, Hammersmith, on the memorial stone of its designer William Tierney Clarke, who completed work on the bridge in 1827. Clarke's stone bridge had two brick piers, above which stood two towers with arched entrances in the Tuscan style. Eight chains were strung from these towers to hold the span of 122 metres in place. The timber deck was divided into a carriageway of six metres and two 1.5-metre footpaths, all narrowing slightly when they passed the towers. It was a lovely and very elegant bridge, the solid towers supporting the graceful, simple sweep of the chains.

As early as the 1850s there was trouble when the bridge rocked after spectators of the University Boat Race rushed from one side of the bridge to the other to watch the rowers pass. The owners were even more alarmed in 1870 when over 11,000 people crowded onto the bridge for the race, and in 1874 they placed 100 spiked bracelets along the chains in order to stop people climbing on them. Later still, access to the bridge was stopped to all traffic two hours before the race in order to avoid damage and overcrowding. By the 1870s concerns were expressed that Hammersmith Bridge was not strong enough to support the weight of the increasingly heavy traffic, even at normal times. The low-slung carriageway also presented problems for shipping, as steamers with tall funnels had to wait for the river level to drop with the retreating tide to limbo under it. In 1884 a temporary bridge was put across the river and work started on a new crossing.

Sir Joseph Bazalgette designed the replacement bridge which opened in 1887 and is best described as enjoyably nuts, and certainly more flouncy (if not bouncy) than the earlier suspension bridge. The centre section between the uprights is 122 metres and the bridge as a whole is 250 metres long with an eight-metre-wide carriageway. The framework of the towers and crossbeams is clad in ornamental cast-iron casings to give the appearance of arches. Decorated iron blocks, supporting the walkway, sit on squat Corinthian columns in the river and the metal appears to cascade down from the towers in fluid, almost organic, waves whose bulk contrast nicely with the far svelter middle section. The bridge is currently painted green and gold and as a whole is a series of flowing swirls, elegant sweeping lines and embossed emblems (sniffily referred to as 'little Frenchy pavilion tops' by the very peeved architectural modernist, Sir Nikolaus Pevsner). 'Harry Potter pavilion tops' might have been a better description as the green scales down their sides are suggestive of a huge dragon.

It is the combination of styles that annoyed Pevsner: layers of what he would see as unnecessary decoration cover the bridge, turning the metropolitan crossing into a cavalier riot of medieval associations. There are even some pleasing heraldic designs on the towers: the Royal Arms of the UK, Guildford (the county town of Surrey), Colchester (the county town of Essex), and the crests of the county of Kent and the cities of London and Westminster. These are interesting because the bridge was built in a period of flux in the government of London – the London County Council had not been set up yet, so there is no crest for it, or the local councils – and, in the number of crests depicted, the Metropolitan Board of Works seemed to be laying claim to a vast area. Eventually parts of Kent, Surrey and Middlesex (which is strangely left off the bridge) did form the LCC, along with Westminster and London, but the remit never extended as far as Colchester or Guildford. By the end of the century, Hammersmith would be the most westerly borough of the LCC, but in 1827 the *Journal of the Franklin Institute* in their review of the original bridge described a 'suspension bridge now erecting at Hammersmith, *near* London'.

On the approaches to the current Hammersmith Bridge there are signs that say it's a weak bridge with a weight restriction of 7.5 tonnes. It has been plagued with problems: the decking was renewed in 1976 and, after cables failed in 1983, these too were replaced. Workmen accidentally caused a fire in 2000 and there were also a couple of attempts to bomb the bridge in 1996 and 2000. In the space of thirty years Hammersmith Bridge was closed, repaired, opened, bombed, closed, re-repaired, bombed again and eventually re-opened. It is perhaps ironic that the landmark bridge in a district named after a local forge should have such deep-seated metal-work problems. During much of the 1990s the bridge was closed to traffic (except for bikes) and served quite happily as London's most ornate foot crossing.

Midway between Hammersmith Bridge and Barnes Railway Bridge to the west is the boundary of the old LCC and this is now the point where the borough of Hammersmith and Fulham stops and Hounslow starts. Just over the border is the

Chiswick Eyot (or Iot, Ait, Eot – Scrabble players take note). This river island provides a sanctuary on the river for young fish and other animals to breed, while waterfowl are amply catered for to the south-east of the bridge by the impressive London Wetlands Centre. This forty-hectare wildlife reserve, opened in 2000 by the Wildfowl and Wetlands Trust, was artificially constructed on the site of the former Barn Elms reservoir and waterworks.

As recently as 1960 the idea of a nature reserve on the Thames might have been considered a joke in poor taste, as the river in London was then so badly polluted it was declared biologically dead. By the early nineteenth century much of the river's wildlife had been destroyed by pollution and habitat loss. There was a revival of fish life towards the end of that century when sewage was removed from the Thames, but pesticides and chemical pollution replaced it as a threat to wildlife and resulted in a second 'death' of the river by the 1950s. German bombing of sewage plants in the 1940s didn't help either.

In 1960 the Greater London Council and the Thames Water Authority embarked on a long-term project to clean up the river. This has worked so well that in 2004 thousands of fish and fry were reported to have died in the Barnes Reach in a single day. The fact that this relatively small stretch of river contained so many fish to kill could ironically be seen as a measure of progress. The fish were killed by an oxygen deficit known as a 'sag' which was caused by the release of sewage during a heavy storm. To help counteract a potential 'sag' two ships pump up to thirty tons of oxygen a day into the water, and there are two shoreside plants that inject hydrogen peroxide into the river.

Water quality is monitored for the Environment Agency by the *Thames Guardian*, and it was onto the deck of this boat that a red-bellied piranha was dropped by a (no

SHE'S IN A SITUATION

Hammersmith Bridge

doubt) stunned seagull in 2003. There are 121 species of fish found in the river, as well as members of other larger aquatic species such as Marilyn the seal, who cavorts further upstream near Teddington, and the porpoises who visit the City possibly to check on fish stocks. There are also 350 species of invertebrates, from leeches to crustaceans, which help support growing populations of water birds and mammals. One of the most successful creatures in the Thames is the Chinese mitten crab, which has made its home for decades along the riverbanks, including the Chiswick Eyot. It is believed that mitten crabs arrived in the Thames after a boat returning from China discharged water into the river. The first recorded sighting was in 1935. The crab was named because of the large amount of hair that grows on its claws, making its pincers resemble mittens. A fully-grown adult can comfortably straddle a dinner plate, which is appropriate as they are a delicacy in China. Eating them may prove the best means of curbing their numbers because, like grey squirrels, they threaten native species. This may have been what John Prescott had in mind when he noted the crab's resemblance to his political rival in the Blair administration, Peter Mandelson.

Even though the wildlife situation is now much better than it was in the mid-twentieth century, threats still remain. Three thousand tons of rubbish is removed from the Thames in London each year and 396 sewage works along the length of the river clean the sewage and return the treated water to the river. It has been estimated that, on average, a drop of rain falling into the Thames at its source will have been drunk by eight people before it reaches the sea. The results of all this recycling are yet another set of pollutants whose effects are less predictable than raw sewage or industrial waste because they are trickier to filter out. These include very low levels of the female sex hormone oestrogen, that may be having a detrimental effect on male fertility by contaminating London's drinking water. The six million or so males who get their tap water from the Thames but are not worried by this development may well have been calmed by sipping traces of the anti-depressant Prozac from the same glass.

Hammersmith residents have other more obvious sources of concern: they are under the Heathrow flight path, and they endure constant traffic jams and the frequent (and related) shutting of their bridge. Three of those closures over the years have been due to the IRA, who were behind the bombs of 1996 and 2000, as well as an attempt in 1939. Why the Fenians pick on Hammersmith Bridge is anyone's guess (though they did blow up a paint store on Chelsea Bridge in 1971 and planned to torpedo Rennie's London Bridge in 1884) – unless it's a plot to unsettle the regulars at the Dove, where James Thompson is said to have composed 'Rule Britannia'.

In 1939 the bridge was saved by the prompt action of ladies' hairdresser Maurice Childs, who was walking across the bridge on the east side at one a.m., and spotted a suitcase. He opened it up, noticed it was smoking then very calmly threw it over the parapet before heading off to call the police. As he reached the phone booth he heard a dull boom and saw a shower of water shooting up over the parapet. This was followed by a second explosion on the west side of the bridge. Two Irishmen were arrested later that night on Putney Bridge and subsequently convicted for the bombing.

The second bomb damaged two suspension rods and some chain links, but the two together might well have wrecked the bridge completely, so the heroic actions of Maurice saved the day. Two decades earlier another brave man, Lieutenant Wood of the RAF, was crossing the bridge when he heard a scream and saw a woman disappearing from the parapet. He dived in after her and managed to pull Margaret Paxton from the river. She was taken to Fulham Infirmary where she recovered. In the days that followed Wood received dozens of letters from admiring women, many proposing marriage. One cheekily said that 'no woman was good enough' for the hero but, if one was, could it be her? Sadly the lieutenant was not able to take up any of the offers as he died two weeks later from lockjaw resulting from a head wound he received from the incident.

There is a plaque in his honour, though none for Maurice and certainly not one for the herbert who in June 2000 threatened to jump from the bridge because he couldn't have a new pair of trainers. Once again the bridge was closed while quick-witted police officers and a generous local sports store provided the eighteen-year-old with a pair of pumps. This is a fairly shocking indictment of young people today when compared to the brave men who risked their lives for the bridge and for others. Though it's unrecorded whether Margaret Paxton was jumping in because she couldn't get a new pair of Mary Janes.

The bridge remains open today despite terrorists (there are CCTV cameras to spot them), traffic (weight restrictions are in place), daleks emerging from the Thames (this was only a film), and even the terrifying ordeal of the scathing words of art critic Brian Sewell. In 1997 he wrote that the bridge was 'a monument of low technology rooted in the Iron Age' and called for a new bridge and the removal of the old one for pedestrian use somewhere else on the river.

Hammersmith Bridge is one of the nicest in London to stroll over and even provides benches halfway across from which to sit and admire the river. The tree-lined views of today are a definite improvement on the Victorian era, when much of the area consisted of factories, mills and various refineries. This is a peculiar echo of

William Morris's *News from Nowhere*, written in 1890. In the book the lead character wakes one morning in Hammersmith on the edge of the city in what to him is the far future and finds the water clean, people fishing and 'the smoke-vomiting chimneys gone'. It would appear that the future has arrived and we can all now enjoy the faux-rural atmosphere of what was once the edge of the Victorian metropolis.

SHE'S IN A SITUATION

Hammersmith Bridge

THIS IS CHARING CROSS

Hungerford Bridges

The original Hungerford Bridge got its name from Sir Edward Hungerford, who in 1682 built a market, where the bridge meets the north shore, on the site of his family home, which had burnt down a few years previously. The market sold only fruit and vegetables until 1833, when it was redeveloped to include meat and fish, and in 1851 a bazaar and art gallery were added. Soon afterwards another fire destroyed the complex, which was subsequently redeveloped (in 1860) as Charing Cross Station. Its name derives from the old English word 'cierran', meaning to turn or bend, reflecting the curve of the river at this point. The cross is a reference to one of the twelve crosses that were placed by Edward I in honour of his wife, Queen Eleanor, to mark the spots where her coffin was rested on its way to Westminster Abbey. The original Charing Cross was made of Caen stone and stood where the statue of Charles I currently is. It was taken down in 1647, and its replacement, erected in 1863, can be seen in the station forecourt.

The official title of the recently built foot crossings here is the Golden Jubilee Bridges, but most people still refer to them as Hungerford and the railway bridge as Charing Cross. As confusing as the names is the question of whether this bridge or bridges are one structure, or two, or three. Since 2002 there has certainly been a physical separation of the railway and foot crossings, but all use the same solid abutments and piers, built for the original Hungerford foot crossing that linked the market to the (then) industrial Lambeth south bank. The four-metre-wide and 415-metre-long footbridge was designed by Isambard Kingdom Brunel and was completed in 1845 after four years' work. It was a very attractive suspension bridge with the platform supported by four chains on rollers that ran through the upper part of the towers. The distinctive Italianate towers were 17.5 metres above the road, and the whole structure was built in brickwork and cement on the natural bed of the river, without using piles.

Within the first twelve hours of its opening, over 80,000 people paid the halfpenny toll to cross it, and 2,000,000 people a year used the specially designed landing stages on the piers that were rented out to steamship companies as moorings. Financially the bridge did very well for its investors, particularly after the opening of Waterloo Station in 1848 significantly increased foot traffic. Subsequently, the Hungerford and Lambeth Suspension Footbridge Company changed its name to the racier Charing Cross Bridge Company (CCBC) and sought powers to convert the bridge to vehicle use as well.

The railways that had brought CCBC so much wealth then threatened to take it away when the South Eastern Railway (which had purchased the Hungerford market site after the 1854 fire) decided to extend its railway line from London Bridge to Charing Cross. Fearing a loss of income from people no longer disembarking at (what is now) Waterloo East and walking over the water, the company sold the bridge to SER and received compensation for loss of future tolls. The footbridge was then closed: one argument used by the railway company for this was that no one would want to walk over a smelly Thames when, presumably, they could sit in a railway carriage over the river instead.

Brunel was not pleased by this outcome. He died in 1859, the same year that his beautiful bridge was partially demolished. The man charged with replacing his bridge, Sir John Hawkshaw, resident engineer of the railway company, was at the time working on the completion of the Clifton Suspension Bridge in Bristol which, ironically, was also designed by Brunel. Hawkshaw incorporated the chains and other suspension elements of the Hungerford Bridge into the Clifton Bridge in 1864 as a tribute to Brunel – and as a canny means of recycling the valuable metal.

The Charing Cross Railway Bridge (also completed in 1864) has, in addition to Brunel's original piers and abutments, extra cylinders of cast iron sunk into the bed to support the whole structure. The superstructure of each of the forty-seven-metre spans consists of two main girders, to the outer sides of which were suspended crossgirders for carrying a couple of two-metre-wide footpaths on either side of the railway track. This maintained pedestrian access, though when the railway bridge was widened in 1882, the upstream footpath was overlaid with track. Nearly a hundred years later, in 1981, the railway bridge's iron girders were replaced by steel, but Brunel's brick piers still support the bridge.

In 1951 a temporary walkway was put in place upstream of the bridge to carry the crowds visiting the Festival of Britain on the south bank. Once the celebrations were over this was removed, leaving again just the downriver foot crossing. This was an ugly, some might say trollish, bridge that linked to the rebuilt Charing Cross Station through a series of gloomy urine-smelling walkways and passed through a small market selling 'collectibles' on the easterly approach to the station. Walking on it was often a cheerless journey, particularly when made alone at night, and the only cinematic use of Hungerford Bridge is aptly in the edgy paranoid thriller, *Defence of the Realm*. Hungerford was also where, in June 1999, two young men were beaten unconscious and thrown over the railings into the Thames by a gang of six youths. One of the men was rescued but the other drowned.

In 2002 two new Golden Jubilee footbridges opened on either side of the railway bridge. They are spacious, airy, removed from the tracks and also relatively puddle-free. Unfortunately they are not trouble-free: a series of late-night attacks in November 2004, on and near the bridges, by homophobic thugs resulted in the death of one man and serious injuries to several others. It seems that words from Ford Maddox Ford's poem 'Antwerp' written in 1918 are still eerily appropriate:

This is Charing Cross; it is past one of the clock;
There is very little light.

There is so much pain.

As Charing Cross is the official geographic centre of London and Waterloo Station is just to the south, the footbridges are very heavily used and there are strong opinions about the crossings here, specifically about the old Hungerford footbridge and its replacement by the new ones. The debate goes to the heart of how the city is perceived by its residents and visitors. To some, London is a grimy, dangerous and exciting city which needs a few dark places in order to continue being something unique, and the old bridge, with its scary twisted iron and dirty paint, certainly provided this. It also offered arguably the best views from any bridge, whereas the sights downriver from the new crossings are not as good and the bridges themselves interfere with, some suggest ruin, the outlook across the Thames. The former juxtaposition of splendid sights downriver with a metal fence and rushing trains behind it on the other side provided a vital urban combination of beauty and squalor. The sanitization of this experience, and loss of the views, are what some residents miss about the old crossing.

For many others the narrow, frequently puddle-laden bridge that hugged the railway bridge was a cramped, damp and unpleasant walk rife with pigeons and reeking of old tramp. There was barely time to stop and look at views at rush hour because the jostling crowd trying to get to and from work or to the theatres meant at best a hurried glimpse before being swept along. The new bridges in contrast can be strolled over comfortably and are accessible to the whole community (a wheelchair would just about have blocked the whole of the old bridge). Each one is, at 4.7 metres, over twice as wide as the older foot crossing and they provide views upriver that have been unavailable (except to railway maintenance workers) since the Festival of Britain.

The need for wider walkways had been recognized by Westminster City Council, which organized a competition for the Hungerford crossings in 1996. The winning team was made up of engineers WSP Group and architects Lifschutz Davidson, and the project itself started in 2000. Construction was complicated by the need to keep the existing railway bridge operating without interruptions. Other problems included the Bakerloo Line tunnels being only a metre or so below the riverbed and (apparently) the potential danger of unexploded Second World War bombs lurking in the Thames after all these years. When money started to run out the Greater London Authority stepped in with a grant to complete the project.

The footbridges are 320 metres long, with the decks suspended from sets of cable stay rods radiating outwards. The two crossings are separate but attached to the railway bridge through outward-leaning pylons. These are secured in place with steel collars fitted around (although not supported by) the pillars of the railway bridge. The white-painted metal struts and stays resemble masts, and evoke a busy Thames filled with shipping. And whether intentionally or not, they hide Hawkshaw's steel girder railway bridge. A reviewer in *The Times* wrote that this might be taken as Brunel's revenge. Certainly the architects did have plans to make more of Brunel's original work, with a transverse connection between the two new bridges passing beneath the railway and through the Surrey pier standing out in the river, but these were abandoned on cost grounds.

Hostile commentators see the bridges as overly complex, a parade of architectural gymnastics parasitically attached to a fine, functioning bridge. They argue that they lack the drama of a genuine suspension bridge, yet ape some of the characteristics. Others find that the spikes point in somewhat random and ungainly directions. But the majority accept the bridges as being better than their predecessor. It has to be said, though, that the walkway through to Charing Cross Station still carries the scent of a urinal.

Crossing the new bridges today it is clear that Brunel's north pier is several metres inland from where it once was in the river and that his bridge crossed a much broader Thames. Just along from the pier, near the upstream footbridge, is a bust of Joseph Bazalgette, who built the embankments, and a plaque on the wall that says, 'Flumini vincula posuit' ('He placed chains on the river'). Brunel's more prominent statue is quite near too, on the Victoria Embankment towards Waterloo Bridge. There have been weirs and embankments at various points on the Thames since Roman times, but it was the Victorians who really created the current shape of the riverbank and built the city we know today. The two individuals who campaigned most strongly for the embankments, albeit for different reasons, were Frederick Trench and John Martin. The former was driven by a need to protect the city from the ravages of nature, whereas the latter wished to protect nature from the pollution of the city. But Bazalgette was the man who actually got them built.

The embankments control the flow of the Thames and literally set its course in stone, and they also add twenty-one hectares of extra land to the increasingly built-up city. They enclose the new sewage scheme but also provide space for the District Line underground and a new road along the river between Westminster and Blackfriars bridges. Part of the genius of Bazalgette lay in the way he built space for gas, electricity and other conduits into his original plans.

The Albert Embankment was completed in 1869 and runs from Vauxhall Bridge to Westminster on the south bank. Without it MI6 would not have its nice new headquarters at Vauxhall and St Thomas' Hospital would not stand where it does near Westminster Bridge. The Victoria Embankment was completed a year later. There was some argument about the road, as the government of the day wanted to build offices on the reclaimed land and use the proceeds to abolish income tax. These plans were scuppered by publishing heir and statesman William Henry Smith Jr, who campaigned for the building of a road and parkland instead. Bazalgette eased the traffic congestion by creating a new road system which fed the extra carriageways along the embankments and provided decent space for walkers. He also strengthened many of the bridges to cope with the increased (and toll-free) traffic flow in the 1870s and 1880s.

By the 1920s there was fresh concern about the bridges' abilities to cope with the increased weight of motorized traffic and a royal commission was set up in 1926 to look into London's strategic traffic needs. A series of crossings were proposed, including one near Chiswick Eyot and a second at Dorset Wharf in the west, as well as a Charing Cross road bridge, with foot crossings slightly downriver of the current ones. This was to be a double-decker bridge with a road to carry cars built on steel girders above railway tracks and, once completed, it would have meant the demolition of the old crossing. In the end Herbert Morrison and the London County Council chose to replace Waterloo Bridge rather than build a new one at Charing Cross.

CROSS RIVER TRAFFIC

Some of the commission's recommendations were carried out, including the broadening and strengthening of several bridges and the construction of new Lambeth, Wandsworth and Chelsea bridges. Since then the focus has shifted from road to pedestrian use and the foot crossings at Hungerford are part of a broader strategy to create a pedestrian-friendly London, with fewer cars, both north and south of the river. An estimated seven million people use the Golden Jubilee Bridges annually. By using them to avoid the roadway on the northern embankment, it is now possible to stroll to the Thames from Chinatown, via Trafalgar Square, across to the equally foot-friendly South Bank, and barely have to cross a road.

I DON'T WANT TO GO TO

Chelsea Bridge

The name Chelsea comes from Celchyth, meaning a landing place for chalk or limestone, the latter no doubt coming in handy when the suburb's palaces and great houses were erected in the Tudor period. Later, in 1681, King Charles II founded the Chelsea Hospital on land that had formerly housed a college of theology and a prisoner of war camp. Christopher Wren completed the hospital in 1692 as a home for veteran soldiers. These 'Chelsea Pensioners' with their scarlet uniforms have been there ever since. The grounds now also host the National Army Museum and, across the road, the Chelsea Barracks. Ranelagh Pleasure Gardens were opened next door in 1742, and when these closed in 1805, they were absorbed by the hospital but continue to offer pleasure to gardeners once a year during the Chelsea Flower Show. Life in Chelsea is not all a bed of roses and Vincent Bayes, in the *Lady's Realm* in 1904, described Chelsea as being 'miasmic, sunless, foggy and pestilential'. The artist James Whistler was more direct: 'dank, dank, dank' was how he summarized the area.

The first Chelsea (or Victoria, as it was once known) suspension bridge was completed in 1858 after seven years' work. The architect Thomas Page made sure that the piers were solidly sunk ten metres down in the clay of the riverbed to avoid problems with the tide. Four cast-iron towers filled with concrete and topped by an elegant lantern were supported on these piers. Intricate castings along the cross-girders gave the bridge the appearance of a Gothic archway, which was matched by the rather fancy toll booths. The old bridge was an ornate beauty in wrought iron and built so that people from the north shore could enjoy the new Battersea Park and set an example to the peasants to the south.

Both the bridge, one of only two paid for directly from public funds, and the park it towers over were part of a social engineering programme designed to gentrify the badlands of this part of Battersea. The area was formerly host to ad hoc encampments, squatters, illegal racing and other sporting events. These centred around a pub known as the Red House which was described in one newspaper article as 'a place out of hell that surpassed Sodom and Gomorrah in abomination'. The charging of tolls (to defray construction costs) and the fact that the bridge was only lit on nights that Queen Victoria slept in London meant that it wasn't used as heavily as initially predicted, although the bridge was always free to cross on Sundays and public holidays. The bridge was 14.3 metres wide and looked like a more robust version of the Albert Bridge soon to be built just up river. This similarity might be due to fact that some of the original ideas for the Albert Bridge were transferred to the Chelsea site after that crossing was delayed by the building of the Chelsea and Battersea Park embankments.

These embankments were finished in 1874 and were the last pieces of engineering by Sir Joseph Bazalgette before he turned his attention to London's bridges. In the late 1870s, after the privately funded bridges had been bought out by the Metropolitan Board of Works, they were thoroughly checked for structural defects before they were made toll-free. In 1876 London, Blackfriars, Southwark and Westminster were already free to cross, but the remaining nine bridges open to road traffic still charged. Waterloo was the first to be liberated in 1878: a huge crowd surged from the Surrey side at the sound of a gun. The following year Lambeth, Vauxhall, Albert and Battersea bridges were freed in a series of elaborate ceremonies, and at Chelsea a parade of Chelsea Pensioners marched over to Battersea Park. In 1879, Wandsworth, Putney and Hammersmith bridges were the last to be rid of tolls.

The verdant appearance of the riverside around both ends of the Chelsea Bridge somewhat hides the fact that many people live around it. The population of Battersea as a whole rose from 3000 in 1801 to 107,000 in 1881 and it has recently been added to again by large-scale developments. The one on the north lies over the old Grosvenor Canal, which was built in 1823 and was formerly run by the Chelsea Waterworks (the first water company in London to have the manners to filter Thames water for human consumption). Grosvenor Dock is also the site where the eastern branch of the River Westbourne flows into the Thames. The large red-brick building downriver from there is Dolphin Square, which could be best described as a housing association for lords and MPs. It was opened in 1937 and contains 1250 flats and, unlike most housing estates, offers room service and tight-lipped discretion.

On the south bank is a series of huge new developments including the Chelsea Bridge Wharf, which is linked to Battersea Park by a small tunnel underneath the bridge. This is part of an ongoing scheme to have a path running all the way along the river embankments on the south side, though obstacles remain around the industrial area near the former Battersea Power Station.

Battersea Park was laid out by Sir James Pennethorne behind sturdy embankments using 150,000 tons of earth shipped upriver from the development of the Surrey Docks. It was opened in 1853 even though the landscaping was not finished until 1864. Battersea retains many typically Victorian park features including a carriage drive, a formal avenue, a lake, flower gardens and shrubberies. And of course a giant Buddha – for what Victorian park is complete without one of those? Actually this was a twentieth-century addition to the park, along with statues by Henry Moore and Barbara Hepworth. In the 1890s female bicyclists shocked locals by riding through the park displaying their ankles, and the first football match ever played under Football Association rules took place there in 1863.

That was the year that the Metropolitan Board of Works made the first modifications to Chelsea Bridge. The bridge continued to need repairs throughout its life, including the fixing of additional chains to strengthen it in 1880, and in 1922 the London County Council removed the pinnacles on the towers for safety reasons. Use of the old bridge almost doubled between 1914 and 1929 to over 12,600 vehicles a day, and the Royal Commission of 1926 was adamant (despite strong public opposition) that a wider and stronger bridge was required. Lack of finance delayed the start of construction until the 1930s, when the government decided that it was better to have men working on the bridge than idly claiming the dole. Material for the replacement crossing was deliberately drawn from across the Empire, with asphalt from Trinidad, wood from Canada, and stone from a variety of regions. And the new bridge was officially opened in 1937 by the Canadian Prime Minister, Mackenzie King.

The self-anchoring steel suspension bridge was built out from either shore, and large central sections, with a combined span of ninety-one metres, lay on barges in the Thames, ready to be put into place. When the shore sections were ready, the middle parts were towed upstream underneath the bridge on a high tide. As the water receded, they were dropped onto jacks that had been specially built to receive them, then hoisted into place and fixed to the rest of the bridge. The twenty-five-metre-wide roadway was then suspended using thirty-seven galvanized steel wires. This honest and robust design was overseen by London County Council architect G. Topham Forrest and the work was completed by E.P. Wheeler.

Despite the straightforward design the present crossing is visually a very striking, if odd, combination of Thunderbird One and seaside pier. The three supports to the towers rest on long, flat piers, and the towers themselves resemble rockets ready for take-off. The bridge manages to be simultaneously dainty yet tough. It has five sets of street lights on either side of the bridge and smaller bulbs fixed into the swooping metal supports. The overall effect of the lighting is pleasing, quite subtle and, best of all, it's on every night, not just when the monarch is sleeping in London. Today the bridge is painted mostly white with a red trim and greyish blue along the balustrades, but in the early 1970s, when it featured more red, questions were raised in *The Times*'s letters page as to why Chelsea Bridge was painted in Arsenal's colours.

There is a series of interesting heraldic designs on the four tall turrets at either end of the bridge. Beneath a golden galleon are two shields (one facing north, the other south) with different symbols on each. Among the motifs are the crests of

Middlesex and other counties around London, and on the south side a series of doves holding olive branches. Bikers now meet peacefully under these doves every Friday evening and have done so for decades. Things were not always so tranquil: on 17 October 1970, two loose alliances of Hell's Angels (the Essex and Chelsea Nomads versus the Road Rats, Nightingales, Windsor Angels and Jokers) had a violent confrontation on the bridge. This resulted in the fatal shooting of one of the Nomads, which brought the feuding to an end, at least on Chelsea Bridge.

Just along from the Chelsea Bridge, by Battersea Power Station, is a series of ten rail crossings that have the appearance of a single crossing. The Grosvenor or Victoria Bridge was built in several stages. Sir John Fowler completed the downriver part in 1860, making it the first railway bridge across the Thames, and the upriver part was finished in 1866 by Sir Charles Fox. These were added to in 1907 to accommodate two more lines of track, bringing the total width to 54.25 metres. Between 1963 and 1967 the whole structure was replaced to carry ten lines of track, each in effect a separate bridge, but part of one complete structure. One of the ideas behind the railway crossing was to connect a posh bit of London directly with the channel ports so that the good people of Victoria could have a nice quick holiday.

The tracks and traffic run both ways and many continentals have made their homes in Chelsea which has a long tradition of welcoming the prosperous. By the 1960s Chelsea was a key part of Swinging London and it is one of the most famous of all London suburbs, with a social cachet and a global brand. These were not enough to hold Suzy Kendall, who, in the film *Up the Junction*, crosses Chelsea Bridge the other way towards Battersea. She was ahead of her time because now more and more people are moving south. It would appear that the bridge has finally succeeded in the goal of gentrifying Battersea: as the King's Road dies, the Northcote

Road thrives, with an en-masse decampment of small specialist shops and artisans from one to the other. Chelsea's charms appear to have been lost on Elvis Costello too, who in the 1970s recorded 'I Don't Want to go to Chelsea', a song that describes a place of transient fashion and jaded sexuality – that's probably a step up though from pestilential and dank.

THERE'S A GRAIN OF SAND IN LAMBETH

Lambeth Bridge

The site where Lambeth Bridge currently stands was a fording point of the Thames long before a bridge was erected. This is where the original Roman Watling Street crossed the river, following an older British track that ran from Dover to Chester and beyond. The Romans settled briefly in Westminster before basing themselves further east but during the Saxon and early Norman period the area became the centre for both the monarchy and the church. This is also the spot where King Cnut, who was the first to build a palace at Westminster, allegedly commanded the tide to retreat.

Most people preferred to take a boat rather than trying to boss the waters or, in the Middle Ages, opting for the horse ferry which was owned by the Archbishop of Canterbury. The horse ferry operated from a jetty near Lambeth Palace until the building of Westminster Bridge just downriver in 1750 made it obsolete. According to all historical sources it was an extremely unpleasant means of crossing the Thames. The language of the ferrymen was foul even by the sweary standards of the river, and figures as diverse as Oliver Cromwell, James I and Nicholas Hawksmoor all wrote strongly worded complaints to the Archbishop of Canterbury after ferries carrying them and their belongings capsized.

Lambeth was also the place where James II threw down the great seal of office as he fled across the river from William of Orange in 1688, believing that without its symbolic authority William would be unable to govern. The seal was found and William ruled until his untimely death in 1702, when his horse stumbled over a molehill on the banks of the Thames near Hampton Court. James's supporters (the Jacobites) ever after toasted the good health of the mole – 'the little gentleman in black velvet' – in honour of the event.

Historical mentions from the fourteenth century of a 'great bridge at Lambeth' are actually referring to a giant landing stage, used on ceremonial occasions by royalty. Permission to build a real bridge between Lambeth and Westminster was first sought from Parliament in 1664. After Westminster Bridge was built, there was not the same pressing need for a crossing, but an Act was passed anyway in 1809 authorizing its construction. Even then, insufficient funds meant that the bill eventually lapsed and three later bids, two in 1828 and one in 1836, failed. It was not until 1860 that the Lambeth Bridge Company finally succeeding in obtaining another Act and sufficient funds to build the bridge, with the growing population of Lambeth and the resulting overcrowding of Westminster and Vauxhall bridges crucial to their eventual success.

The first Lambeth Bridge linked Church Street in Lambeth to Market Street in Westminster. It was designed by P.W. Barlow and opened in 1862. In honour of the old means of getting across, Market Street was later renamed Horseferry Road. The original suspension bridge was 253 metres long with two ten-metre towers in the river and shorter ones on the abutments. Three platforms of equal length were supported by suspension cables hanging from these towers and by heavy crossbeams along the piers. The bridge was apparently a sturdy affair, able to hold 800 tons at a time, and in its early days it raised a good deal of money in tolls. It was less popular with horse-drawn users who found the approaches too steep, and they started to claim that the structure was unsafe. In 1887 the rumours became reality when engineers discovered that one of the piers was beginning to tilt, and major repair work had to be carried out. In 1905 a weight restriction of 2.5 tons was imposed, and by the First World War vehicles were banned completely. Pedestrian access was also regulated, but unfortunately there is no historical record as to if or how stouter strollers were banned.

Parliament blocked London County Council's plans for a new fifteen-metre-wide bridge in 1912, but approved a larger replacement in 1924 and also the funding to improve the approach roads. In 1929 work started on the current five-span bridge, designed by Sir George Humphreys, with architectural support from G. Topham Forrest and Sir Reginald Blomfield. It is made of steel and reinforced concrete with polished granite facings, and it still has quite a curve to it. In the raised sections above the piers there are seats from which to admire the river and panorama of St Thomas' Hospital, Parliament and Westminster Bridge. The bridge is painted a mixture of chocolate brown, red and black. The red reflects the colour of the benches in the nearby House of Lords. Ornate lamp standards and a nice bit of lattice work were later added to the parapets. On each of the piers there are double lamps supported by a granite upright, and at intervals along the balustrades there are single lights on a black lattice support. At the approach to the bridge the lamp standards are particularly unusual – blue lamps supported by a strange fish motif and topped with a crown.

Although more ornamental, the fish sculpture is similar to those on the rest of the Albert and Victoria embankments which are known as the sturgeons lampstands. The fish resemble ornamental carp but are actually stylized Romanesque representations of dolphins. The dolphin was adopted as an emblem of London in the nineteenth century in a deliberate attempt to link London with Ancient Rome, which was the first city to adopt the comedy dolphin motif as symbol of imperial might. Each lampstand has two of the creatures entwined around it, with their heads at the base and tails a metre or so up in the air near the lamp itself. Some date back to the building of the embankments while others along the riverfront were added by the Greater London Council in the 1970s.

Lambeth Bridge is further enlivened by the pineapples festooned on obelisks at its approaches – a colonial-era symbol of friendship and hospitality, which might make this London's most welcoming bridge. It is certainly hospitable from the south side, where the wide approaches embrace travellers and funnel them across. The MI5 headquarters at Thames House and other government buildings on the north bank present a more austere face. One theory about the pineapples is that they were put there in honour of the Tradescants, seventeenth-century gardeners, adventurers and traders credited with introducing various exotic fruits to this country. The Tradescants were buried in a memorial Knot Garden in the churchyard of St-Mary-at-Lambeth, which is now the Museum of Garden History. Others interred there include Captain William Bligh of *Mutiny on the Bounty* fame and William Sealy and his partner Eleanor Coade, whose tomb is made of the Coade stone they invented. Their factory in Lambeth produced this high-quality and very durable artificial stone, made up of a secret mixture of ingredients lost until the twenty-first century, that shrank very little in the firing process and thus allowed accurate reproduction of detail. Legend has it that if the tomb is danced around twelve times as Big Ben strikes midnight, a ghost appears, although no one seems to know of whom or why. Another ghostly event is the re-enactment of the murder of George Villiers, the Duke of Buckingham (1592–1628), who was stabbed to death on 23 August 1628 near where the Lambeth Bridge currently stands. Every year at five a.m. on that date, people gather on the bridge in the hope of seeing the murder restaged in spectral form, and even if George fails to show they can always watch the spooks arriving for work at Thames House.

THERE'S A GRAIN OF SAND IN LAMBETH

Lambeth Bridge

St-Mary-at-Lambeth is mentioned in the *Domesday Book* and is older than the more famous Lambeth Palace next door. The palace had been added to over the centuries until 1829, when part of it collapsed, and Edward Blore designed a new one in mock Gothic style. Some of the older parts still survive, such as the Norman crypt where Anne Boleyn was cross-examined, the thirteenth-century chapel above it and the 1432 Lollards' Tower. The latter derives its name from the Nonconformists who followed John Wycliffe and who were said to have been imprisoned there, even if its actual function was as a water tower. Wat Tyler's rebels sacked the palace during the Peasants' Revolt of 1381, as it was an important symbol of the ruling class. Fortunately for the king and parliament there was no bridge to cross otherwise their palaces might also have been wrecked.

The Thames (and the ability to cross it) has long provided the key to defending

London. William I had to march halfway to Oxford before he could cross and set about attacking from the north bank. Centuries later, holding the river and its banks was vital during the Civil War, and even today the bridges and river are closely watched for terrorist threats, particularly around Lambeth Bridge, with its many important buildings. In another sense the bridges themselves provided a defence for central London by enabling the boundaries of the city to push further afield, thereby protecting the inner core. This is a theme that the poet Stephen Spender picked up during the Blitz, when he took comfort in the fact that nothing as immense as London could possibly be destroyed by mere bombs.

The buildings to the south of Lambeth Bridge exhibit both the haphazard expansion of the area and the mixture of beautiful and ugly that characterizes London's growth. Clues remain just upriver of the bridge as to what the area might have looked like once. On either side of the Fire Brigade pier, arches are cut into the wall on the southern embankment: these are what is left of the docks that served the Lambeth Doulton pottery factory which is an ornately tiled red-brick building that can still be seen in Lambeth High Street and Black Prince Road. The rather attractive lions etched into the embankment with rings through their mouths for moorings now fulfil a decorative rather than practical function. The later twentieth-century building next to the fire brigade headquarters houses the International Maritime Organization, which is responsible for monitoring safety at sea and is fronted by the Seafarers' Memorial statue. On the bridge approach itself is the twenty-first-century Parliament Heights apartment complex.

Eighteenth- and nineteenth-century bridge building provided the means by which London could expand southwards, gobbling up villages and hamlets as the roads to the bridges attracted new settlements. In this way Lambeth, Walworth,

Newington, Clapham and even the marshy area around what is now Waterloo were rapidly transformed from being largely rural to predominantly urban. In 1801 the population of south London represented about fifteen per cent of London's total; today it is nearer a third. Not all residents were happy with these changes; many were shocked by the deprivation that came with urbanization (the term 'dole' for poor relief was coined in Lambeth, named after the alms distributed from Lambeth Palace). William Blake, who lived on Hercules Road in Lambeth, wrote the poem 'London' about the increasingly oppressive and growing city:

I wander thro' each charter'd street,
Near where the charter'd Thames does flow.
And mark in every face I meet
Marks of weakness, marks of woe.

In every cry of every Man,
In every Infant's cry of fear,
In every voice, in every ban,
The mind-forg'd manacles I hear.

Hardly a hymn of praise to the growing metropolis. The theme was picked up in the late nineteenth century in a music-hall song about the stunning vistas to be seen across London, 'If it Wasn't for the 'Ouses in Between'. London's growth, aided by the bridges, was truly astonishing. Births only exceeded deaths for the first time in 1790, so London was clearly being driven by immigration and colonization of previously separate villages. Better communications from suburb to centre and the deteriorating state of the Thames meant that richer residents began to move away from the river, a process that has only been reversed in recent decades. Many MPs remained in Lambeth as, because the area is within the sound of the division bells, it is possible for them to make it to the chamber and vote within the fifteen minutes permitted. Unions and pressure groups also had their headquarters in the vacinity because of its proximity to Parliament. These include, strangely in view of their name, the Countryside Alliance, though it rarely put in an appearance at the Country Show held every summer in Lambeth's Brockwell Park. Having a country show in the heart of the city is a good reminder of the area's history. Lambeth derives its name from 'lamb's hythe', signifying a harbour where livestock and other farm produce were landed.

CROSS RIVER TRAFFIC

Lambeth was the first, in 1965, of London's crossings to be tunnelled beneath to provide pedestrian access along the embankment: it is perhaps a shame that none of the area's history is told on the blank walls of the underpass on the south side. Lambeth is an important crossing that provides a link between three of the traditional seats of power in British society, though the bridge itself honours another much more recent force. The coats of arms of the old London County Council are sculpted on the piers, below which two granite arms curve down to the top of the cutwaters. This was the first bridge built by the LCC which by the 1920s was democratically and confidently managing the affairs of London without recourse to king or church.

MISTY MORNING

Albert Bridge

Architectural critic Phillip Howard described the Albert suspension bridge as being in a 'fussy Victorian grand manner with intricate cat's-cradles of cable and curious little pagodas that crown the supports'. A late-Victorian commentator wrote that the bridge had no architectural beauty, and Geoffrey Phillips, in his book *Thames Crossings*, rather dismissively suggests that Albert Bridge resembles a seaside pier rather than a metropolitan crossing. To describe the bridge as ornate is an understatement. Camp might be a better adjective for a bridge painted pink and lit with what appear to be fairy lights. On the uprights are shades of blue and green with yellow and white trim for support. If areas dictate the kind of bridge they get, then this Mr Kipling confection of a bridge is ideal for linking this particular part of Battersea with Oakley Street in Chelsea.

In 1842 the Commissioners for Woods and Forests recommended construction of an embankment at Chelsea to free up land and allow a more efficient road network to be built. They also suggested that the embankment could link to a new bridge that should be built to supplant the old timber one at Battersea. Work started in 1864 but by the time the Albert Bridge was completed, the new Chelsea Bridge was already in place downriver, and plans to knock down Battersea Bridge had been revised to include a replacement crossing there. This meant that when the three-span iron bridge, designed by Rowland M. Ordish, was finally opened in 1873, it became one of three crossings on this reach of the Thames.

CROSS RIVER TRAFFIC

The Albert Bridge has four twenty-one-metre-high iron towers supported on cast-iron cylinders filled with concrete which support the structure. It is kept stiff by rigid metal bands radiating from the towers, which resemble minarets crossed with a spray of stays. Viewed from the river, the bridge resembles the opening credits of a Walt Disney film. It has been described as a wedding cake and a surrealist fantasy, and in 1967 the *Chelsea News*, under the headline 'New Dress for Trembling Old Lady', described positively the 'gay sepia and white effect around [the] portholes'. It is the sort of bridge that one might associate with the Danube in Prague, which is where Ordish built the similar Franz Joseph Bridge in 1868. This was demolished in 1949 and Ordish's London counterpart could have gone the same way.

As part of the deal to get Albert Bridge built, the Albert Bridge Company, confident that there was still money to be made from toll bridges in west London, agreed to take responsibility for the old Battersea Bridge. This extra expense drained resources away from constructing better approach roads to Albert Bridge itself, which meant that fewer people than anticipated used the crossing from the start. Five years after its opening Albert Bridge, along with Battersea Bridge, was bought by the Metropolitan Board of Works, which immediately inspected the crossing for faults. Several were found and over the next decade the Board added new chains, reinforced the structural supports and laid down a new deck, but plans to demolish it altogether were put on hold and it survived into the twentieth century. As a result of the alterations, the bridge more closely resembles a conventional suspension bridge, and even after the modifications a five-ton weight limit had to be imposed.

On the embankment footway to the north of the bridge there is a nude female figure by F. Derwent Wood but no statue of Prince Albert himself, though it could be that the bridge itself is monument enough. Also on the north shore is a statue by David Wynne called *Boy with a Dolphin* that nicely twins Albert Bridge with the similarly flamboyant Tower Bridge, which has the same artist's *Girl with a Dolphin* statue by it, at St Katharine's Dock. Both bridges have, at various times, been described as overly decorative Victorian follies and were dragged into a broader nineteenth-century debate between those builders who were driven by ideals of honest architecture and those who preferred a more design-based approach. These arguments were much fiercer over the construction of Tower Bridge because of its strategic location, but Albert Bridge was strongly criticized too. Tower Bridge has gone on to become an icon of London and a tourist attraction in its own right whereas the Albert is popular with the natives, but a surprise discovery to many visitors.

The bridge's north bank has always been fairly affluent, and because of this there is little modern housing there, but to the west along the southern walkway is the London headquarters of architect Sir Norman Foster. Foster, along with Richard

Rogers whose own offices are near Hammersmith Bridge, has been responsible for the large-scale redevelopment of London's Victorian riverscape – many of the more iconic structures, and certainly a good number of the twenty-first-century system-built buildings along the Thames banks are Foster's responsibility. Between them he and Rogers have managed to start a more recent debate about architecture on the river. It is pleasingly contradictory that Foster should have chosen to build his office overlooking Albert Bridge's ornate Victorian splendour.

Albert Bridge has not been threatened by either Rogers, Foster or any other twenty-first-century architect, but it spent most of the previous century in virtually constant danger of demolition. The 1926 Royal Commission's report on London Bridges states, 'If and when funds are available, this bridge should be rebuilt and widened to carry four lanes of all classes of traffic.' The depression and war depleted funds and, though the weight limit was reduced to two tons in 1935 and it was closed for repairs in 1951, it was not until the late 1950s that serious proposals were put forward to replace the bridge. These were fought successfully by an alliance of local residents and fans of the bridge, led by the poet John Betjeman.

If it had been possible for all the former inhabitants of Cheyne Walk just to the north of Albert Bridge to organize a pressure group it would have made a formidable gathering. George Eliot lived at number 4 and Lloyd George at number 10. Dante Gabriel Rossetti lived at number 16 in a building formerly known as the Queen's House, built in 1719 by John Witt. Rossetti moved in during 1862 and kept a small zoo there. The noise made by the peacocks caused such annoyance to his neighbours that subsequent leases on property in the area specifically forbade the keeping of the birds. Henry James lived and died in nearby Carlyle Mansions, and Marc and Isambard Kingdom Brunel lived at number 98. Other residents of the walk or its environs have included Sir Thomas More, Oscar Wilde, Hans Sloane, Thomas Carlyle and Hilaire Belloc. James Whistler, Joseph Turner and John Sargent also lived in the area, reflecting Chelsea's strong links with the arts. This connection has recently faded as younger artists have been priced out, and in 2004 the Chelsea College of Art relocated, arguably removing any remaining spark of creativity on the King's Road.

Some artistic types still make use of the bridge, and its fey qualities have been exploited by those intent on making London a chocolate-box (office) backdrop for romantic comedies. Tower and Westminster are the most heavily filmed bridges and Waterloo the only one that plays the central role in a movie; Albert Bridge specializes in the fluffier output of the film industry. It wouldn't do to blame this doxy of a bridge for its use in a few witless films, including *Absolute Beginners*, *Maybe Baby* and *Sliding Doors*, but perhaps the Greater London Council or today's Greater London Authority should have thought of selling it off to a film studio. This would have brought in some cash to build a proper solid metropolitan structure that Mr Howard and the other critics could be proud of.

In 1970 GLC engineers predicted that the bridge would remain standing for only another twenty years and suggested that it be closed to traffic immediately prior to being destroyed on public safety grounds. The Albert Bridge Group successfully lobbied to have it repaired and, in response, the GLC suggested strengthening the bridge and either closing it entirely to traffic or operating a limited service for cars running south to north in the morning and north to south in the evening. Then the *Architectural Review*, in conjunction with the GLC, came up with a beautiful plan to convert the bridge into a park by pedestrianizing and landscaping it. This prompted an odd alliance headed by Diana Dors and the Royal Automobile Club which successfully argued that it should be kept open for cars as well as pedestrians. Eventually two concrete piers were built under the main span and a new, even lighter, deck was laid, enabling the bridge to be reopened to some traffic in 1973.

The braces mean that a crossing which was lucky to survive the nineteenth century has made it into the twenty-first, despite the fact that its fragility had been recognized very early on. Signs set onto the toll booths that remain at either end of the bridge instruct soldiers to break their march when crossing it, in case the rhythmic pattern of their step sets up vibrations that would damage the crossing. Today a greater threat comes from incontinent hounds who cannot hang on till Battersea Park, as their urine is rotting the wood underneath the bridge. Perhaps it's just as well *101 Dalmatians* wasn't filmed on it.

CROSS RIVER TRAFFIC

Battersea Park is naturally a huge attraction for all canines and those who love them because it contains the legendary brown dog. This is a replacement of an earlier statue that Battersea Council commissioned in 1906 to show their support for the anti-vivisection cause. The monument to the unknown mongrel outraged sections of the medical community, who promised to destroy it, and it ended up with a police guard. In 1911 a Conservative council was elected in Battersea that removed the poor mutt from his plinth and threw him into the Thames. It wasn't until the 1980s that a new statue was cast and placed in the park.

The Albert Bridge forms part of a nice route for joggers who wish to run a verdant square along the river through the park, back over Chelsea Bridge and along the tree-lined north bank past the Chelsea Physic Garden. This was founded in 1673, as the Apothecaries' Garden, to train apprentices in identifying plants and learning their uses. The proximity to the river created a warmer microclimate, allowing many non-native plants to survive the harsh British winters, helped by an underground heating system up and running by 1684 in some of the greenhouses. The river was also important as a transport route linking the garden to other open spaces, and for receiving deliveries from overseas, and today the garden is (alongside Kew Gardens upriver) a world centre of bio-diversity.

The bridge perched there in its pastel shades stands out in contrast to the greenery around it and looks like an antique leftover from another era. The current colour scheme is part of a 1992 makeover and represents a more varied pattern than the previous golden yellow colour (1981–92) and solid green (1905–81). The lighting system was revamped at the same time and the 4000 bulbs, strung along the bridge, were reset or replaced and the cute original street lights on the pathway were renovated. Albert Bridge is gorgeous in the day but spectacular when lit at night and stands in contrast to some of the more functional crossings along the river. In another position on the Thames its charm would probably not have saved it from being replaced. It owes its survival less to its innate beauty and more to its location and the failure of the Albert Bridge Company to built decent approach roads, because there is no point constructing a wider bridge with such narrow approaches in place to serve it. It is many people's favourite London crossing and perhaps Geoffrey Thomas was not far off the mark when he described it as resembling a seaside pier – it can make you feel like you are on holiday.

WANDLING FREE

Wandsworth Bridge

The Squeeze song 'Cool for Cats' contains a line about inmates of Wandsworth Prison having numbers and names on their shirts. But the lyrics could equally evoke the sight of Chelsea fans in their replica kits herding to and from Stamford Bridge grounds across the Wandsworth Bridge. Their shirts (with numbers and names on them) even match the blue and white of the current crossing. The Wandsworth Bridge was completed in 1940 by Sir T. Peirson Frank, and replaces an earlier iron lattice bridge of 1873 built by J.H. Tolme. More vehicles cross it than any other London bridge but it is also, without question, the ugliest and least inspiring of all the London crossings. It links Wandsworth town centre on the south side with what was originally an industrial area of Fulham to the north. There was some debate between the local authorities of Wandsworth and Fulham about the final position of the bridge, but the privately owned company that built it settled eventually in favour of the site preferred by Wandsworth Council.

The London Borough of Wandsworth was established in 1965 by merging the former metropolitan district of Battersea with about two-thirds of what then constituted Wandsworth. It is the largest of London's boroughs and has been transformed by gentrification since the 1980s. The young and mobile were attracted by the very low poll tax, relatively cheaper housing and a feeling that they were moving to an area full of kindred spirits. They are in good company historically, as Daniel Defoe, Thomas Hardy, William Makepeace Thackeray and François-Marie Voltaire all once lived in the neighbourhood.

The Wandsworth Company's original plan was to build a suspension bridge similar to the Albert. The two bridges received royal assent on the same day in 1864 and both suffered lengthy delays because of the building of the Thames embankments. They were also the last bridges in London speculatively built by private companies for profit. Developers saw the opportunity for revenues from tolls and profits from increasing land prices around the bridges as London expanded. The hold-ups over Wandsworth Bridge resulted in rising costs and by 1870 the company was forced to seek an alternative to the original proposal, particularly after the designer (Rowland Ordish) became involved in other projects.

Julian Tolme's simpler design had four iron-based piers backed up by brick abutments, and the company applied for more capital and for the right to construct extra piers in the river rather than the planned two. The five-span crossing supported iron girders on which was laid a ten-metre-wide wooden deck. It was opened in 1873 by Colonel Hogg, and Mrs Hammond of the Spread Eagle pub laid on the celebratory buffet. The cold banquet was a fine affair, but the bridge was a bit of a dog's dinner, with its ungainly structure and ill-matched materials. It is very hard to find anyone who had a good word for it, with terms such as 'fright' and 'eyesore' being most frequently used to describe it.

The bridge's appearance was not the only problem. The company that built it had agreed to compensate the owners of nearby crossings for projected loss of trade, yet failed to raise much money from tolls. Partly this was because the sulky local council on the north side did not adequately maintain (or even drain) the roads that led to the crossing. This severely reduced vehicle traffic across the bridge, which was the largest source of income, as carts paid sixpence to cross and pedestrians only a half-penny. The investors then hoped for a large compensation package when the Metropolitan Board of Works purchased the bridge in 1878, but unfortunately the amount of recompense was based on the poor toll revenues received. Furthermore, when they contested their award in court, a clutch of engineers and bridge experts all testified to the shoddy condition of the crossing and the poor workmanship involved.

Maybe less care was taken over this bridge because, as *Wandsworth Borough News* has claimed, Wandsworth was traditionally not really orientated towards the Thames and until recently has not made the most of its situation on this reach. As much of this stretch of the river was built up during the Victorian era when the river was dying, industrial and smelly, perhaps this was wise. Arthur Harry Beavan, writing in *Imperial London*, describes the views across the Thames to Wandsworth as London's 'least comely' and continues:

> The vision is of brick, a curtain of dirty brick, cutting the background of leaden sky in hard uncompromising lines without a curve or bend. Looking hopelessly to the right for some relief from this depressing picture, there is nothing but the commonplace height of Wandsworth crowned with an unsightly church . . . resembling a gigantic skewer of Brobdingnagian proportions.

In 1880 a five-ton weight limit was imposed on the bridge and the decision was taken to replace it as soon as possible, particularly as it was too feeble to carry buses. Strange, then, that it took fifty-five years before a design for the current Wandsworth Bridge was approved. A temporary bridge was put in place while the old crossing was cut up and towed away, and in its place arose Sir Peirson Frank's steel and concrete cantilevered bridge, which was completed in 1940 and widened in the 1970s.

Local residents hoping for a beautiful new crossing must have been very disappointed. There are few concessions to glamour even if the two concrete piers have been roughly cladded with granite and the same stonework is used on the corner pylons. The bridge is more streamlined than its predecessor and is a solid utilitarian structure that would not look out of place crossing the Mersey near Runcorn. It also has a vaguely military air about it, which is perhaps appropriate as it was constructed in the build-up to the Second World War when steel was at a premium, and it was painted a dull greyish blue to camouflage it from the Luftwaffe. Geoffrey Phillips, in *Thames Crossings*, kindly says that it 'adds variety to London's bridges' but only in the sense that a box of chocolates is not complete without those orange fondue confections that no one likes to eat.

Much of the focus of Wandsworth was towards its own river, the Wandle, on the banks of which the business-minded Huguenots once produced hats for Rome's top clergy (despite the fact that they were Protestants driven out of France because of their religion). These migrants were typical of the industriousness of the good people of the municipality as a whole, and were also evidence of the tolerance of the locals towards new arrivals, a fact further exemplified by Wandsworth's proud boast of having elected the UK's first black mayor, John Archer, as long ago as 1913.

Ever receptive to migrants, Wandsworth got its name from a Saxon tribal leader known as Wendel, and is a corruption of Wendelsworth, meaning 'Wendel's enclosure'. The river Wandle, which empties into the Thames just upriver from the bridge, derives its name from the same source. The Wandle has been cleaned up over the past few decades and is distinctive in that it is above ground when it enters the Thames, and throughout most of its course. The majority of central London tributaries are buried, though the Westbourne (at Sloane Square) and the Tyburn (at Baker Street) can be seen through pipes at those underground stations. The Wandsworth Bridge is close to other feeder rivers including the Falcon, whose name is derived from the crest of the Lords of Battersea Manor, and which in turn shares its name with numerous pubs in the area, but has no connection to the beautifully named Drunken Bridge, which once crossed a stream near Garrett Lane, and like the stream is no more.

There are thirteen 'rivers' that enter the Thames in the London County Council area. Some of these are buried streams which were once quite substantial rivers, but which now can only be spotted at low tide, when the drainage pipes that conduct them are exposed. These subterranean watercourses are revealed through the names of streets or areas: Effra Road, Stamford Brook and Fleet Street. The Walbrook, an important watercourse in Roman London, now exists in darkness from its source in Moorfields, just an echo of its former self and a cause of rheumatism for those who live above its channel. The Tyburn manages to pass through many of London's most famous addresses unseen and its name is associated with the place where London's criminals were once hanged. The Westbourne, whose waters help form the Serpentine in Hyde Park, enters the Thames near Chelsea Bridge. Other waterways that snake through London include the Ravensbourne, Lea, Beverley Brook, Earl's Sluice and Neckinger, which runs into the main river at Bermondsey. There are also a number of legendary watercourses that have been truly lost such as the Langbourne near Fenchurch Street, the Cranbourne that allegedly runs underneath Soho, and Parr's Ditch in Fulham.

In the past these rivers acted as defensive fixtures and land boundaries; they were a source of drinking water and power as well as being valuable for water-intensive industries such as tanning and metalwork. More obviously they were used for irrigation and drainage and supported industrial fisheries as well as recreational angling. For centuries they were also transport routes: King Cnut sailed up the Effra, as did Elizabeth I five hundred years later to visit Walter Raleigh in Loughborough, no doubt in order to pick up something dodgy to smoke.

During heavy rains the old rivers sometimes still flood, as the watercourses fill up rapidly due to run-off rainwater from the surrounding concrete. The stench of the trapped waterways has a harmful effect on human health. One thing Oliver Cromwell and Charles II had in common was being afflicted by the 'London Ague', which was a form of malaria prevalent near the rivers. Even if obvious water-borne diseases like cholera or Weil's disease are ignored, rates of tuberculosis, influenza and even allergies are higher along the old river courses. And G.W. Lambert discovered in 1960 that there is a strong correlation between the streams and reported paranormal events. Up to seventy-five per cent of reported hauntings in some areas occur near the buried rivers, and certainly sudden cold or clammy spots, draughts and strange sounds could be caused by changes of water flow.

Covering over rivers seems a shade disrespectful when in the past they were worshipped and the bridges that spanned them were places of dread, wonder and sacrifice. In southern France and northern Italy a loose organization of knights, clergy and artisans, the Fratres Pontifices or Frères Pontifes (Brothers of the Bridges) had a monopoly on bridge building for centuries. The Pope or 'sovereign pontiff' derives his title from the chief priest of pagan Rome, the Pontifex Maximus, or great bridge-builder, himself a link between the divine and the mundane. In other religions bridges have to be crossed to reach paradise after death and the unworthy are turned back or made to wait endlessly on the bridge.

Concepts such as hell and eternity should be easy to grasp for anyone who has ever been stuck in a traffic jam on Wandsworth Bridge. It is appropriate that Wandsworth Bridge marks the point where the speed limit on the Thames drops to eight knots (upriver). The chief reason for the speed restriction is that upriver of Wandsworth the Thames is used more frequently by rowing clubs, and rowers would suffer if motorized boats were allowed to whizz past them or create huge swells. Below Wandsworth Bridge there is no specific limit – safe speed depends upon the local conditions and the important factor is that boats present no danger to persons or other river traffic.

The upsurge in use of the Thames and the Wandle for recreation is beginning to make the area more attractive. Crossing the bridge to the south, having left an uninspiring cauldron of roads in Fulham, is still a fairly bleak experience, but the views are improving. Much of the industrial area that caused Beavan such distress is being redeveloped for housing, with new projects stretching all the way towards

Battersea Bridge. The southern bank is already a long way advanced around Wandsworth Bridge and, with the borough being re-orientated towards the Thames, hopefully it will only be a question of time before the whole waterfront (and the bridge with it) is tarted up.

In general this gentrification is a positive thing and has led to many improvements, but this is not universally the case. Near Tower Bridge some recently arrived residents complained about moorings on the river in front of their flats and the sight and sounds of the boats despite the fact that the mooring rights go back centuries. A similar situation occurred near Wandsworth Bridge in 2003 when a group of art students, sketching the river from the pathway, were asked to move on by an admittedly embarrassed concierge. The concierge was acting on behalf of a resident of one of the flash new blocks who felt it within his rights to move people along because they were interfering with the expensive views he'd paid for. A fair conclusion from this might be that it is time for some of the newer residents to acquire the tolerance for which the borough of Wandsworth is justifiably famous.

A TOWER IN THE HEART OF LONDON

Tower Bridge

Something as fruity as Tower Bridge would never be made today – even though it ably fulfilled its remit to ease congestion by providing a route over the Thames east of London Bridge, without blocking river traffic access to the busy Upper Pool docks that were once known as the larder of London. The idea of dressing up a steel bridge as a medieval castle was criticized at the time it was built, but conservationists were keen that the Tower of London should not be overshadowed by something too modern-looking. Tower Bridge is meant to complement the much older Tower of London and the fact that many visitors assume the bridge has been around for longer than just over a century is further proof of the success of the design.

Architects have condemned Tower Bridge's lack of purity, even if it is a good example of a building blending into existing surroundings.

The first proposals were made in 1870, but the winning design and necessary parliamentary approval did not come until 1885. Between 1874 and 1885 thirty petitions from various public bodies were brought before the authorities, urging either the widening of London Bridge or the building of a new bridge. All the competing architects had to address the fact that the Upper Pool still contained large ships discharging their cargos. Any bridge would have to be high (or flexible) enough to allow the bigger trading vessels through. Among the fifty schemes considered were low-level bridges with swing openings of various kinds, high-level crossings with inclined approaches or with lifts at either end, and a rolling bridge with sections that moved across a series of supports. A subway was built in 1870, but John Keith's plans for a giant under-river arcade of shops was vetoed. A free ferry option such as the one introduced with great success by Joseph Bazalgette at Woolwich was also dropped.

Bazalgette's plan for the bridge – a giant parabolic arch – was the last serious rival to the eventual winner, by City Engineer Sir Horace Jones. The deciding issue was whether the bridge should be large, bold and modern or respect the existing architecture of the area. Jones had something of an advantage as he was on the team of judges, and in 1878 was asked to offer a critique of Bazalgette's bid on behalf of the judging panel. In the end the committee went with Jones's plan for two short suspension bridges from either shore with larger central towers operating a chain drawbridge.

Some didn't want a bridge at all, including the parishioners of a church on the north bank who objected because they thought it would bring undesirables into their area. They had a point as some of the worst slums in London and the city's grimmest residents did live to the south-east of the proposed bridge at Jacob's Island in Bermondsey. This was an area borrowed by Charles Dickens (who lived in Southwark as a child, and who was no stranger to deprivation) for Bill Sykes's lair, described as:

> the filthiest, the strangest, the most extraordinary of the many localities that are hidden in London, wholly unknown, even by name, to the great mass of its inhabitants.

Queen Victoria was also unamused by the prospect of the bridge, fearing that a crossing so close to the Tower would threaten the defence of London. The Tower had been an armoury for centuries and was perceived as an important strategic fixture. The idea of a bridge leading straight to it struck her as abhorrent, and in a letter she also rejected claims that the bridge would improve the area.

> to those who say the bridge will increase the defensive strength of the Tower and improve the beauty and historical associations of the place. All I can say is bosh!

Interesting choice of words from a boche queen who eventually backed down because the issue was so politically sensitive. The architectural press at the time did not care about political niceties or the clamour of public and mainstream press support, and was scathing about the plans, describing the bridge as 'architectural gimcrack', 'a monstrous and preposterous architectural sham' and 'a discredit to the generation that created it'.

Compared to some of the alternative proposals, especially the rolling bridge, the rather over-ornate bascule bridge seems perfectly sensible. A bascule bridge (derived from the French word for see-saw) was the idea of the bridge's engineer, John Wolfe Barry. It is a kind of drawbridge that moves on a pivot with a heavy weight at one end counterbalancing the greater length at the other. In 1976 the original 1200-ton bascules (counterbalanced by 422 tons of lead and iron) were replaced by new ones as part of the conversion from steam power to electricity.

Tower Bridge architect Sir Horace Jones died in 1887, one year after work started, so it was George D. Stevenson who finished the bridge and changed Jones's original brick covering to stone, as well as adding some Gothic flourishes. During its construction Tower Bridge was known as the Wonder Bridge because, as well as being the largest, most sophisticated bascule bridge ever built, it also used hydraulic power on a scale never attempted before. Seventy thousand tons of concrete were used to build its supporting piers and 11,000 tons of Scottish steel were required for the framework of the towers and walkways. It also used 22,000 cubic metres of Cornish granite and Portland stone and 20,000 tons of cement. Material was brought from all over the UK and the 31,000,000 bricks alone would be enough to build a wall half a metre high along the Thames from source to sea.

Tower Bridge was constructed in sections and floated downstream on barges from Woolwich. Two major piers were sunk into the riverbed to support the construction: these are 56.3 metres long and 21.3 metres wide, with central areas of 6.5 square metres forming the base of the towers. The main towers have columns 36.5 metres high, which the ten-metre-wide roadway passes through and the two four-metre footpaths go around. The side spans are suspension platforms supported by chains anchored in the rear of the abutments and carried over the entrance arches to the main towers. Here they are joined by rods concealed in the two walkways, providing support for the roadway's lifting mechanism. The foundations are secured against the tide and the columns in each pier are braced, to resist the wind pressure of up to four kilograms per square centimetre. The masonry covering forms effective, as well as attractive, protection for the metal underneath against extreme fluctuations of temperature, among other things.

During its early decades the bridge opened more than 6000 times a year and a staff of 120 was employed, mostly to stoke the furnaces to get the steam up. If you want a decent idea of how busy the Upper Pool and Tower Bridge were up until the 1950s you could do worse than watch Basil Dearden's film *Pool of London*, which shows the docks fully operational. Surprisingly, despite the thousands of openings and closings, there have been no major accidents on the bridge, though in the 1950s a double-decker bus complete with passengers did have to make a jump for it when the bridge opened while it was crossing, and in May 2004 a breakdown in communications led to a ship, the *Lord Nelson*, crashing into the bridge when the roadway was not raised. The roadway itself is flat and this was another positive feature of the design, because it didn't scare or tire the horses as a taller or curved bridge might. London was still a horse-drawn society at the start of the twentieth century, with a quarter of a million working horses in the city in 1900. Humped bridges and the prices of hay and oats were vital concerns then, in the way that petrol costs and the congestion charge (of which Tower Bridge marks a boundary) are today.

Even after it was completed the architectural press were still critical and the editor of the *Builder* magazine refused to show pictures of the finished bridge, preferring to print sketches of the steel substructure. The daily newspapers were wholeheartedly in favour, with *The Times* describing it as 'the greatest engineering triumph of the Victorian age' and the *Illustrated London News* calling it a 'magnificent and most useful structure'. The public, both British and foreign, took to it immediately, and there was a fierce pride, particularly among those who lived near the bridge or worked on it, about the marvel that had been erected. One French visitor kindly noted that 'no other people know how to unite with the same harmonious force the cult of the past, the religion of tradition, to an unchecked love of progress and a lively and insatiable passion for the future'. The Gothic grandeur of the structure, together with the smooth operation of the raised roadway, is something that excites visitors and residents even today. In terms of London iconography it has eclipsed the Tower and other older symbols of London.

Naturally its visibility made the bridge a target for German bombers in the Second World War, but despite their best attempts the closest they came to destroying the bridge was when a flying bomb bounced off the 'cross bar', and the southern approach was damaged when a bomb hit nearby docks. The bridge was painted battleship grey for camouflage at the time – rather than the more upbeat red, white and blue of today – though it's doubtful that this would have been enough to disguise it. Still, it was an improvement on the original chocolate-brown colour scheme. More modern threats, such as traffic, that have plagued the other Victorian bridges, barely affected Tower, but a weight limit of five tons was introduced in January 1975. Traffic flow was speeded up in 1954 when the automatic barriers were introduced to block the road prior to the bridge being raised, replacing the manual system of dragging chains across the roadway.

Today one man pressing a button is all that's required each of the 1000 times or so the bridge opens annually. This is nearly double the number of openings of a decade ago, reflecting an upturn in river use. Tower Bridge takes between three and five minutes to open. The walkways above the bridge were rather superfluous, as most people were prepared to wait and watch the bridge open rather than climb 43.5 metres above the Thames. Others, meanwhile, were impatient to reach the walkways: despite the close attentions of police, diver Benjamin Fuller died in a plunge from the walkways shortly after the bridge opened. In 1910 the walkways were closed to the public because they were only really popular with suicides and hookers, the latter showing their ankles off to fine effect on them. They did not reopen until 1982, after being glassed in.

Underneath the bridge on the north shore there is a gate and steps to the water's edge, that lead to Dead Man's Hole, where bodies were pulled out of the river and stored in a mortuary above. A happier, and more common, sight under the bridges (north and south) are the tourists visiting the area's many attractions, including the Tower itself, which draws in 2.5 million punters a year. Sadly they can no longer spend time at the 250-metre pleasure beach (created in 1934 using 1500 tons of sand) just in front of the Tower as it was closed in 1971. The strand could accommodate 500 people at low tide and rather ironically it was shut on public health grounds, despite the fact that by then the river was less contaminated than at any time since Tower Bridge was built.

As well as enabling tourism, Tower Bridge functions as a doorway to the financial district, which is the centre of London's most lucrative industry. Around a third of the City's workforce are employed in business and allied service industries, up from ten per cent in the 1960s. London still carries on as a trading city, but now instead of goods physically entering London's docks, money enters City institutions before being traded on, or visitors spend money on local goods and services. Young professionals from these industries form the majority of the people who are moving into the former harbour areas, whose population doubled between 1984 and 2004. St Katharine's Dock to the north-east of Tower Bridge was the last quayside constructed near to the city, and one of the earliest to close. In 1969 the Greater London Council decided that the area should become a mix of private and public housing, office accommodation and recreational facilities. It was to be the blueprint for much of the later Docklands redevelopment and also the setting for one of the better British crime films, *The Long Good Friday*, which stars Bob Hoskins as a London villain seeing opportunities in the redevelopment, yet trapped by his own parochialism.

The film is one of the best metaphors for London's reconstruction in the 1980s and 1990s, when the city entered a new global era, but struggled (as Hoskins does) with the myths of its past and its new position in the world. From 1965 onwards the inner London dockyards began to close because of greater competition from foreign and British container ports nearer the sea. Other factors contributing to this were their relative lack of efficiency, poor infrastructure and outdated equipment, so that by 1983 the last of them (Millwall, West India and Royal Albert) were shut. The impact of the closure of the docks, at least in terms of direct employment, has been somewhat overstated. In the post-war years in inner London, there were just over 60,000 registered dockers, a total which fell to 23,000 by 1969 and 12,000 by 1973. At their peak, adding in allied trades, perhaps 100,000 people were employed on the wharfside, but this is considerably less than the half-million light industrial jobs that disappeared from London between 1961 and 1974.

Regeneration is not an exclusively late-twentieth-century concept. In the immediate post-war period there was a scheme to upgrade Tower Bridge by surrounding it with glass. The plan, by W.F.C. Holden, included 19,000 square metres of office and housing space, but was thrown out because Tower Bridge, only fifty years old, was already a heritage icon. After the closure of the docks, the southern approach to Tower Bridge became very rundown, but the Upper and Lower Pools have since been comprehensively redeveloped with offices, shops, housing and City Hall, the new Greater London Authority building. Butlers Wharf, completed in 1873, once the largest warehouse complex on the Thames, is now best known for its restaurants. The distinctive iron bridges in Shad Thames, which helped in the movement of goods from warehouse to warehouse, have become decorative features for the designer flats of the area, which appropriately also boasts a museum dedicated to design.

The Design Museum serves in a way to highlight the striking success of nineteenth-century design that is Tower Bridge. This crossing has become a symbol of both London and the Victorian age, even if some regarded it as a bit of a folly. Tower Bridge is still the most easterly of the London crossings, but as the concept of central London has shifted east, it is now closer to the heart of the metropolis and the gateway to the City of London. In many ways it is the perfect emblem for London as an ever-changing trading city able to transform itself in the face of altered circumstances. When one bridge closes another opens.

A TOWER IN THE HEART OF LONDON

Tower Bridge

A DASHING BLADE

Millennium Bridge

London is peculiar. The city waits a hundred years for a new bridge, then three (the Millennium and two at Hungerford) appear almost at once. After they collectively won a competition in 1996, Foster and Partners, with collaboration from sculptor Anthony Caro, designed the Millennium Bridge, which was built by engineers Ove Arup and Partners. It is a very shallow suspension bridge with two Y-shaped armatures supporting eight cables that run along the sides of the four-metre-wide deck. Steel transverse arms clamp onto the cables at eight-metre intervals to support the walkway itself. The cables never rise more than 2.3 metres above the path, making for an oxymoronically flat (or nearly flat) suspension bridge that allows crossing pedestrians better views of the river and surrounding buildings. The aim was 'a slender arc across the water', by day a 'thin ribbon of steel and aluminium' and by night 'a glowing blade of light'. The centre span does have something of this quality and, seen from above, the outriding sprays swoop down to the side of the bridge like running-boards on a car. The best views of it are actually obtained from below, by approaching along the north bank where there is an odd illusion of great height, as if the bridge is dangling there above the buildings.

The Millennium Bridge opened to huge excitement and large crowds, only to close promptly because it was too popular. Organizers had estimated that fewer than 10,000 people would cross it on the first day in June 2000, but they ended up with more than 100,000. This volume of people tramping across it on the opening weekend set up a rhythm that caused the bridge to move seventy millimetres on its support. Although the 'nausea-inducing ride' as some described it (in fact it was merely a feeling of mild intoxication, with your feet going slightly awry, and not unpleasant at all) never threatened to undermine its structural integrity, the bridge's excessive movement forced designers and engineers back to their drawing-boards. They discovered that as few as 500 to 600 people in marching formation were enough to set the bridge swaying. One mathematician even came up with a formula to express it: $F = K \times V$, where F is the sideways force exerted on the bridge by pedestrians, V is the sideways velocity (or movement) of the bridge, and K is a mathematical constant. F creates V as people's steps lock in to each other, resulting in the wobble.

The Millennium was not the only bridge to suffer 'wobbles'. Old London Bridge had all manner of problems, from fires to bits falling off, and one of the themes of the associated nursery rhyme is that the bridge had had to be repeatedly built up using different materials. There was the Wembleyesque series of fiascos that affected

Westminster Bridge, arguments arose over who should rebuild Blackfriars Bridge, and Vauxhall Bridge went through several design mutations during construction. Further west the old timber crossings at Battersea and Fulham made the river hazardous for shipping. All the bridges built between 1750 and 1850 had to be replaced because they had become unsafe or could not cope with the demands of modern traffic. Those built in the later Victorian period have survived, admittedly in the case of the Albert and the Hammersmith after major support work.

Many of these problems were eventually dealt with by Sir Joseph Bazalgette and the Metropolitan Board of Works, which also took care of another outstanding problem by buying the bridges off the companies that built them. This was a relief to investors: far less money was raised by most bridges than expected, even from the relatively prosperous ones like Vauxhall and Waterloo, and very few had made anything like a decent income. Beautiful though many of the older bridges were, only those who owned land on either side of them did well for themselves. Southwark Bridge Company and the Albert Bridge Company, for example, were practically insolvent before the crossings even opened, and what traffic there was (mostly pedestrian on the toll crossings) was slowed by the toll booths. The buyouts were effectively a case of local government stepping in where private enterprise had failed.

In February 2002 the Bridge House Estates Trust took over responsibility for the Millennium Bridge, which means the bridge is in very safe hands. The Trustee of the Estates is the Corporation of London, and there is no financial support from the government or any other public source: bridges increase trade, and that is the basis of the Trust's riches. If more people are crossing the bridges, more money is coming into the City and the value of land around the bridges increases, so the Trust gets greater revenue from the rents on that land. The pedestrian focus of the bridge has also had the ironic effect of boosting the trade of the old enemy of the bridges, the watermen. The large number of walkers it helps bring to Bankside has increased river traffic, including an 'art boat' service linking the two Tate galleries at Bankside and Millbank, with a stop-off at the London Eye. It could be said that the Millennium Bridge has helped to lighten the loads of many, even if it couldn't bear heavy ones itself.

The bridge was closed to the public until the wobble was cured by the installation of a series of shock absorbers, or dampers, beneath the bridge (so as not to spoil the elegant lines and sweep). One nice nod was made to the past when workers correcting the bridge's sway used the traditional warning method of bales of hay hung from the bridge's support. This was to alert any passing shipping that work was taking place above, to signal to those on board to watch out for falling spanners. When the Millennium Bridge was re-opened to the public in February 2002 restrictions were initially introduced to limit the number of people allowed on the bridge at any one time. For the re-opening, no doubt in the spirit of scientific enquiry, a newspaper put a woman in a bikini carrying two plates of jelly on the bridge, to see if anything jiggled.

The Millennium Bridge links the tourist belt around St Paul's Cathedral in the north with the Globe Theatre and Tate Modern to the south. It has good disabled access, with a ramp on the southern side and one of London's more interesting lifts on the northern bank that runs from the footpath to the bridge at a forty-five-degree angle. No doubt Christopher Wren would have found the crossing very handy when building St Paul's, as he used to live in a house between the new Globe and the Tate Modern. The bridge would also have been useful for the punters going to the theatres, bear-pits and brothels that Bankside formerly specialized in. Actors can today be seen on the bridge being filmed, crossing to work at the Globe, or conducting walking tours when 'resting' from their main profession. The sight of bears crossing to work would however be a novel and eye-catching spectacle, though perhaps little more so than a Native American in full head-dress blessing the waters from the bridge, who those lucky enough to be crossing on 21 September 2004 were treated to. Then again if any waters were, historically, in need of ritual benediction by a strange man in feathers then the Thames would certainly qualify.

The Tate Modern had five million visitors in the year after it opened in May 2000, and a constant flow of millions thereafter. But some of the unofficial attractions in the area, such as Big Chief Joseph above, can be equally diverting. The Budgie Man leads the field, though, in a show that is part *Blue Peter*, largely avian and completely bonkers. With the aid of several household items and a small group of volunteers, he sets his birds a series of tasks whereby they move through obstacles to get from one perch to another. He does this while singing a self-penned promotional song interminably. It's fantastic entertainment and, apparently, unique to Bankside – unlike many of the rest of the menagerie of unaccomplished street painters, irksome statue impersonators (it's not like we lack the real thing), twisted metalworkers, and the rest of the tedious international buskers' roll-call that can be seen from Sydney to Seattle.

These ad hoc minstrels compete with the series of official festivals that run from May to September and celebrate aspects of London life, its communities and the river Thames itself. Throughout the summer the space along the river is also host to a wide range of theatre, comedy, music and dance, which has brought to life an area of London that was derelict for many years. Some of these attractions can be seen from the new bridge, which was clearly designed with the sights in mind. The views are comfortable too, as a good deal of thought was put into protecting pedestrians from the wind that whips along the river.

The bridge has also made the northern walkways along the river easier to reach and has connected them to the open esplanade to the south, which has brought into sharper focus the issue of access to the river. Some campaigners are concerned that the 'privatisation' of the riverfront will reduce the public's right to walk along the Thames, even if no such 'right' existed in the past. Joseph Conrad noticed on his first visit to London how inaccessible the river was, in comparison to other ports, due to the wall of buildings right up to its banks. The docks themselves were a fairly secure working environment and strangers stood out in the close-knit Thameside communities. After they closed, people seeking an interesting, if potentially dangerous, stroll along the river, around the rotting buildings and rusting machinery, were free to roam (twenty-four-hour attack dogs permitting), though relatively few did. As recently as the late 1990s it was possible to take a Bankside bike ride from Tower Bridge to Vauxhall in the evening and barely see a soul. Today it is virtually impossible to do because of the crowds, so it would seem that there are now no problems with the public reaching part of the waterfront at least.

If earlier proposals had been acted on, though, there would be no Millennium footbridge to provide this access. In 1911, J. Lewis put forward plans for a St Paul's Bridge across the river at the same point, in order to link the Elephant and Castle in the south to Angel Islington in the north by tram and road. The idea was revived in the 1920s and it was hoped that this bridge would boost trade and transport links, helping Southwark's development and property prices. Similar plans for tram and road links were behind the proposed Ludgate Bridge just upriver of the Millennium Bridge site.

The Thames is littered with abandoned projects, from Robert Vazie 's and Richard Trevithick's 1808 Limehouse Tunnel to a range of bridges planned for west London in the 1970s. Proposals are currently well advanced for several crossings east of Tower Bridge. One of these is a fresh tunnel and two others are railway crossings. There are also plans for road and foot bridges, one from Greenwich to Silvertown and the second from Beckton to Thamesmead (Thames Gateway Bridge), which has received planning permission and been referred to a public enquiry. The current blueprint is for the Thames Gateway Bridge to carry a six-lane dual carriageway connecting the A406 North Circular Road with the A206 in Thamesmead. Ironically it is in exactly the same position as the proposed East London River Crossing rejected in the mid 1980s. The idea is for road, cycle and pedestrian access, but also for some of the lanes of the bridge to be dedicated to public transport use only.

The Thames Gateway Bridge is part of a broader focus on the eastern river. There are projects for a river bus facility and residential development at Convoys Wharf in Deptford, to encourage more river traffic and improve access to central London from the east. Use of the river as a means of transport for commuters is rising again after a period of decline, with new Thames Clipper services introduced in 2004, and journeys by river are often speedier than any other means. It takes roughly fifteen minutes by riverbus from the Tate Modern pier to Canary Wharf, for example. Leisure use of the Thames is expanding, from the facilities on its banks to rowing, sailing or pleasure cruises on the river itself. It is worth noting as well that central London's Thames still handles more material (by tonnage) annually than all of London's airports combined, or the equivalent of 400,000 lorries every year on the streets of London.

The Millennium Bridge crosses a very different Thames to the one of a hundred years ago when Tower Bridge was built. The river is cleaner, and yet its waterfront is more heavily used than at any time in the city's history. London appears to be rearranging itself, with the river once again at its heart. The old Bankside is booming and, as in the pre-industrial city, places of entertainment line the shores. The Thames is now an easily crossed, beautiful feature rather than a barrier or defensive fixture. What better place from which to view the rejuvenated city than the newest of the bridges that links two of the oldest parts of the metropolis together?

APPENDICES

LENGTH, CURVE AND ABUTMENTS

Brief bridge facts

Foot and road crossings

Albert Bridge Designed by Rowland M. Ordish and opened in 1873. Albert is a suspension bridge made of iron, wood and concrete, 216 metres long, 12 metres wide and 4.9 metres above the river.

Battersea Bridge Designed by Sir Joseph Bazalgette and opened in 1890. Battersea is a five-arch bridge made of iron and granite, 203 metres long, 16.7 metres wide and 5.5 metres above the river. It replaces an earlier crossing from 1771 built by Henry Holland.

Blackfriars Bridge Designed by Joseph Cubitt and H. Carr and opened in 1869. Blackfriars is a five-arch bridge made of iron and granite, 281 metres long, 32.4 metres wide (after widening in 1910), and the tops of the five arches are 7.1 metres above the water. It replaces an earlier crossing from 1769 built by Robert Mylne.

Chelsea Bridge Designed by G. Topham Forrest and E.P. Wheeler and opened in 1937. Chelsea is a self-anchoring steel suspension bridge resting on stone piers, 200 metres long, 25 metres wide and 6.6 metres above the water. It replaces an earlier crossing from 1858 built by Thomas Page.

Hammersmith Bridge Designed by Sir Joseph Bazalgette and opened in 1887. Hammersmith is an iron and stone suspension bridge, 250 metres long, 12 metres wide and 3.7 metres above the water. It replaced an earlier suspension bridge from 1827 built by William Tierney Clarke.

Hungerford Bridges Designed by Lifschutz Davidson architects and opened in 2002. The twin steel Golden Jubilee Bridges are 320 metres long, 4.7 metres wide and 7 metres above the water. They run either side of Charing Cross Railway Bridge (completed in 1864) and replace a walkway that ran along the downriver side of the railway bridge, which in turn replaced a footbridge built in 1845 by Isambard Kingdom Brunel.

Lambeth Bridge Designed by Sir George Humphreys and opened in 1932. Lambeth is a five-arch bridge made of steel and reinforced concrete with polished granite facings, 236 metres long, 18 metres wide and 6.5 metres above the Thames. It replaces an earlier suspension bridge from 1862 built by P.W. Barlow.

London Bridge Designed by Mott, Hay and Anderson and opened in 1973. London is a three-span pre-stressed concrete bridge, 300 metres long, 32 metres wide and 8.9 metres above the Thames. It replaces an earlier crossing of 1831 built by John

Rennie, which in turn supplanted Old London Bridge, which had stood for 600 years. This was preceded by a series of wooden Saxon bridges as well as Roman bridges dating to AD 50 and c. AD 120. One of the odder facts about the Roman bridges is that they did not cross all of the Thames, which flowed in a series of small channels hundreds of metres to the south of the current waterfront. These first bridges linked Londonium to a river island from which it was a series of short hops across other islands and marshland to the real southern shore nearer where Borough tube is today.

Millennium Bridge Designed by Foster and Partners and opened in 2000. Millennium is a shallow suspension bridge made of concrete and steel, 330 metres long, 4 metres wide and 8.7 metres over the water.

Putney Bridge Designed by Sir Joseph Bazalgette and opened in 1886. Putney is a five-arch granite bridge, 202 metres long, 25 metres wide (after an operation to broaden it in 1933) and 5.5 metres above the water. It replaces an earlier wooden bridge of 1729.

Southwark Bridge Designed by Sir Ernest George and opened in 1921. Southwark is a five-span bridge of steel resting on stone, 203 metres long, 17 metres wide and 7.4 metres above the water. It replaces an earlier bridge built in 1819 by John Rennie.

Tower Bridge Designed by Sir Horace Jones and opened in 1894. Tower is a bascule bridge of steel covered by stone. The total length of the bridge, including the approach roads, is 700 metres, but the width of the river between the abutments of the bridge on the north and south sides is only 268 metres. The width between its Gothic towers is 61 metres and the bridge is 8.6 metres above the water with its bascules down, increasing to 42.5 metres when they are raised.

Vauxhall Bridge Designed by Sir Alexander Binnie and opened in 1906. Vauxhall is a five-arch bridge of metal resting on stone, 246 metres long, 24 metres wide and 5.6 metres above the river. It replaces an earlier crossing of 1816 built by James Walker.

Wandsworth Bridge Designed by Sir T. Peirson Frank and opened in 1940. Wandsworth is a three-span cantilevered steel and concrete bridge, 197 metres long, 18 metres across (after widening in the 1970s) and 5.8 metres above the water. It replaces an earlier 1873 bridge designed by J.H. Tolme.

Waterloo Bridge Designed by Sir Giles Gilbert Scott and opened in 1942. Waterloo is a five-span crossing made of concrete faced with Portland stone, 381 metres long, 24.3 metres wide and 8.5 metres above the water. It replaces an earlier 1817 bridge designed by John Rennie.

Westminster Bridge Designed by Thomas Page with Sir Charles Barry acting as consultant and opened in 1862. Westminster has seven cast-iron arches supported on granite plinths and is 252 metres long, 24 metres wide and 5.4 metres above the water. It replaces an earlier 1750 bridge designed by Charles Labelye.

Railway bridges

Battersea Built in 1859, the whole crossing is 500 metres long, 279 metres of which is over the river in five 56-metre spans. There are an additional six brick arches on the northern and southern shores. Opened in 1863 the bridge had two lines of track, one for standard gauge and one for the wider gauge of the Great Western Railway system.

Blackfriars The first Blackfriars Railway Bridge was built by Joseph Cubitt and opened in 1864. It was a five-span wrought-iron girder bridge carrying four lines of track to Blackfriars Station and the first railway bridge to cross to the City. In 1886 a second 32-metre-wide bridge was built by W. Mills, assisted by John Wolfe Barry and H.M. Brunel (second son of I.K. Brunel), to connect with the station of St Paul's.

Cannon Street (formerly Alexandra Bridge) Built by Sir John Hackshaw and opened in 1866 this bridge is 278 metres long and 32 metres wide, of which 25 metres consisted of track, with 7 metres for the two footpaths. The upstream path was lost by the end of the nineteenth century, while the downriver one survived the extensive reworking of the bridge during the First World War, but not that of the Second World War.

Charing Cross Originally this was a reworking of Brunel's suspension bridge, which kept the piers and abutments of that crossing. Sir John Hawshaw added extra supports to make a six-span bridge, 567 metres long, that opened in 1864. It was widened for tracks in 1882 by incorporating the upriver footpath into the railway and extending the piers. The bridge was strengthened in 1918, and from the late 1970s to early 1980s the original wrought iron was replaced by steel.

Putney Designed by William Jacomb and completed in 1889. This lattice girder bridge is 301 metres across the river in five spans with two 39-metre spans on the southern shore and one on the northern making the bridge 418 metres in total. A pedestrian footpath runs along the upstream side.

Victoria (Grosvenor) Originally opened in 1860 the bridge was 366 metres long (291 metres over the river) and 12 metres wide. An additional 39-metre-wide bridge was added by 1866, supported on cast-iron cylinders in line with the original piers. In 1907 further lines were added, along with 18.5 more metres of bridge, bringing the total width to 70 metres, with the extra piers made to blend with the original. Between 1963 and 1967 the whole structure was replaced to create ten separate strands of one track each that appear to be one broad bridge.

THEY BUILT THIS CITY

Some engineers and architects

In reality London was actually *built* by thousands of anonymous, often migrant labourers who carried out much of the work for little pay and at great personal risk. However this brief list of biographies gives a little more information on some of the men of vision and influence who have made their mark on London. It also serves to differentiate between the various Rennies, Cubitts, Barrys and so forth.

Sir Charles Barry (1795–1860) travelled widely in his youth before designing churches for the Church Commissioners in London, Brighton and Manchester. In 1824 he started the Royal Institution in Manchester before building the Travellers' Club and the Reform Club in London. He also developed the classical Treasury Building in Whitehall and most famously rebuilt the Palace of Westminster.

Charles Barry Jr (1823–1900) carried on his father's work at Dulwich College where he was, like his father, chief architect. He also rebuilt Burlington House in Piccadilly and the Great Eastern Hotel at Liverpool Street.

Edward Middleton Barry (1830–1880) completed the Palace of Westminster for his father and redesigned the Covent Garden Opera House. Edward also designed the

Charing Cross and Cannon Street hotels, reconstructed the Eleanor Cross and built the Sick Children's Hospital at Great Ormond Street.

Sir John Wolfe Barry (1865–1904) is another member of the Barry clan but he added the Wolfe part to his name in 1898 presumably so he could be easily distinguished from his relatives. He carried on the family fondness for the Gothic in his additions to Tower Bridge, and was responsible for Earl's Court Station, the extension of the Metropolitan District Line, and many other London stations.

Sir Joseph Bazalgette (1819–1891) rebuilt three of the older Thames bridges (Hammersmith, Putney and Battersea) and repaired several more, as well as introducing the free Woolwich ferry. He did more to shape modern London than probably anyone else, overseeing the building of the Thames embankments, roads, park schemes and, most famously, the London-wide sewerage system.

Sir Alexander Binnie (1823–1917) was born in Ladbroke Grove and became the chief engineer for the London County Council, building the first Blackwall Tunnel (1897), the Greenwich foot tunnel (1902) and Vauxhall Bridge (1906), and extending London's sewerage system.

Isambard Kingdom Brunel (1806–1859) aided his father in completing the first Thames Tunnel. Isambard built the Hungerford footbridge in 1845, planned the Clifton suspension bridge over the River Avon and constructed a network of tunnels, bridges and viaducts for the Great Western Railway. He also redesigned or constructed many of Britain's major docks including Bristol, Monkwearmouth, Cardiff and Milford Haven.

Marc Isambard Brunel (1769–1849) was born in France but left for the US in the mid-1790s and became New York City's chief engineer. In 1799 he moved to England, where his work in mass-producing ships' blocks mechanically meant that the British could build ships faster than their rivals (principally the French); this contributed to the victories over Napoleon. Brunel designed machines for sawing and bending timber, boot-making, knitting and printing.*

Joseph Cubitt (1811–1872) was part of a family from Norfolk who turned out an amazing number of civil engineers. He was the son of Sir William Cubitt and assisted his father working for the South Eastern Railway, becoming chief engineer in 1846. He completed several railway lines in the south and in Wales, as well as Weymouth Pier and Blackfriars Bridge.

* Marc Brunel based his revolutionary tunnelling shield on the activities of the common shipworm (*Teredo navalis*), which digs with two shells above its head and excretes a brittle layer behind to support the tunnel it has just created. Brunel could not put his conception into immediate practice as he was in prison until 1818, but he patented his idea on his release. He built a giant shipworm with four-metre corkscrew blades, propelled by hydraulic jacks. As the shield advanced, men followed it and built up the tunnel behind, but after a year of work the Thames broke in, as it did again in 1827 and 1828. This last flooding caused the project to be abandoned, leading *The Times* to label the scheme 'the great bore', and an obituary for the tunnel to appear in Priestley's *Navigable Rivers and Canals*. In 1835 the government provided finance to finish the tunnel, which took another five years and the death of ten workers (mostly through gas poisoning but a couple through decompression).The tunnel was officially opened in 1843 and 50,000 people passed through it on the first day.

Lewis Cubitt (1799–1883) was the brother of Thomas Cubitt and the nephew of Sir William. He is best known as the architect of King's Cross Station and of the Great Northern Hotel, also at King's Cross.

Thomas Cubitt (1788–1855) did not, like most of his family, involve himself with transport projects. He was however one of the earliest, and most important, of the speculative builders who built up large areas of London, including sections of Belgravia, Pimlico, much of Bloomsbury and Kemp Town in Brighton.

Sir William Cubitt (1785–1861) patented self-regulating windmill sails in 1807 and in 1817 he invented the treadmill used in British prisons. From 1826 to 1858 he practised as a civil engineer, working on a number of improvements to the canal network before setting up large sections of the South Eastern Railway, including the 1268-metre Shakespeare Cliff tunnel. He made the biggest noise when blowing up the face of Round Down Cliff using 8165 kilograms of gunpowder, though was more delicate when supervising the construction of the Crystal Palace for the Great Exhibition of 1851.

George Dance the Elder (1695–1768) served as London city surveyor and architect from 1735 until his death. He built Mansion House and a number of churches, as well as St Luke's Hospital, and was the last man to erect significant buildings on Old London Bridge.

George Dance the Younger (1741–1825) took over from his father as city surveyor and architect for London in 1768. His ideas for Blackfriars Bridge and the replacement for Old London Bridge were turned down but he was responsible for rebuilding Newgate Prison (1770) and constructing the church of St Bartholomew-the-Less.

Sir Norman Foster (1935–) founded his Associates in 1967, now known as Foster and Partners, and has designed prestigious buildings across the world. These include the reconstruction of the Reichstag in Berlin, the Great Court at the British Museum and the new Hong Kong international airport. Other buildings in London include the Millennium Tower and the Swiss Re headquarters, aka 'The Gherkin'.

Sir John Hawkshaw (1811–1891) worked for several railway companies as a civil engineer and was responsible for the Charing Cross Railway Bridge and the Severn Tunnel, completed Isambard Brunel's Clifton suspension bridge, and converted Marc Brunel's Thames Tunnel for railway use.

Sir Horace Jones (1819–1887) became the Chief Architect for the City and designed Leadenhall and Smithfield markets, as well as Tower Bridge. He is buried in West Norwood Cemetery.

John Martin (1789–1854) started off as an heraldic painter and his early pictures were exhibited at the Royal Academy in 1812. From the 1820s onward Martin spent most of his time on engineering projects including the embanking of the Thames. Martin believed that the natural world was a repository of a pure goodness that must be separated from man-made corruption – except when it came to recycling man's sewage to benefit nature. He is probably the only person to talk passionately about human waste to a Select Committee in Parliament, in 1838.

Rowland Mason Ordish (1824–1886) is probably most famous for designing the Albert Hall but this architect and engineer was also responsible for the Albert Bridge and St Pancras Station.

Robert Mylne (1734–1811) is particularly remembered for his work in the late eighteenth century when he built the first Blackfriars Bridge and the Hunterian Medical School (now part of the Lyric Theatre). He is buried in St Paul's, where he served for many years as surveyor to the cathedral.

Thomas Page (1803–1877) started his career as assistant engineer to I.K. Brunel on the Thames Tunnel and went on to have his designs chosen for the Thames embankments. In all he built four bridges over the Thames (Chelsea, Westminster, Windsor and the non-tidal Albert Bridge at Datchal); however his designs for Blackfriars and Tower – as well as those spanning the Rhine at Cologne, the Danube at Budapest and the Golden Horn – went unfulfilled.

George Rennie (1791–1866) built a bridge over the Serpentine and was the man who did the calculations for many of his father's and brother's constructions, but his genius was mostly mechanical. He set up an engineering firm that produced machinery for many prestigious projects, including the tunnelling shield for Marc Brunel's Thames Tunnel.

John Rennie (1761–1821) attended Edinburgh University from 1780 to 1783 before beginning work as an engineer. In 1791 he relocated to London. He was responsible for a number of canals, docks and drainage schemes but is best known as a bridge builder. He combined stone with cast-iron to create revolutionary bridge designs, including Leeds, Waterloo and Southwark bridges. His last project was London Bridge, which was completed by his son John. Unlike his son, he refused the knighthood offered to him. He is buried in St Paul's Cathedral.

Sir John Rennie (1794–1874) was the youngest of John Rennie's sons, chosen to complete his father's design for London Bridge and also responsible for dismantling the Old London Bridge. He and his brother George were involved in the construction of George Stephenson's Liverpool and Manchester Railway in 1830.

Sir George Gilbert Scott (1811–1878) was a leading architect of the Gothic Revival and responsible for Albert Memorial and Midland Grand Hotel, St Pancras, among other projects.

George Gilbert Scott Jr (1839–1897) followed the career path of his father and was responsible for a few country houses, some college buildings in Cambridge and the Catholic cathedral in Norwich.

Sir Giles Gilbert Scott (1880–1960) continued the Scott design dynasty and can claim the red telephone box, Waterloo Bridge, Battersea and Bankside power stations, and Liverpool's Anglican cathedral among his successes.

Frederick Trench (1775–1859) started his career as a soldier and worked his way up to becoming a general. Although in later life he became a politician and gifted amateur architect his attitudes to nature appear to have been strongly formed while fighting in Holland. There he noted the prosperous healthy farmland around the structured dykes and the disease-ridden marshes away from them. From this he concluded that the forces of nature must be contained or they will become a source of disease and destruction.

SELECT BIBLIOGRAPHY

Additional and invaluable information was provided personally by Bob Jefferies of the MSU, who was a fine source of river lore (and law), and Sean Griffiths of FAT, whose views on bridges and architecture were a tremendous help.

Bridge by Bridge Through London: Thames from Tower Bridge to Teddington by Tony Waters, Pryor Publications, 1989

Bridges Over the Thames by Ruth Mindell and Jonathan Mindell, Blandford Press, 1985

A Dictionary of London Place-Names by David Mills, Oxford Paperbacks, 2001

'Downward Mobility: Victorian Women, Suicide, and London's "Bridge of Sighs"' by L.J. Nicoletti, *http://homepages.gold.ac.uk/london-journal/march2004/nicoletti.html*

Geographers' London Atlas, Geographers' A–Z Map Company, 2001

The Great Stink of London: Sir Joseph Bazalgette and the Cleansing of the Victorian Metropolis by Stephen Halliday, Sutton Publishing, 2001

A History of London by Stephen Inwood, Macmillan, 2000

Location London by Mark Adams, New Holland, 2003

London: The Biography by Peter Ackroyd, Vintage, 2001

London Bridge: A Visual History, Peter Jackson Historical Publications, 2002

The London Encyclopaedia edited by Benn Weinreb and Christopher Hibbert, Macmillan, 1983

London's Thames by Gavin Weightman, John Murray, 2004
The Lost Rivers of London: A Study of Their Effects Upon London and Londoners, and the Effects of London and Londoners on Them by Nicholas Barton, Historical Publications, 1992
Making the Metropolis: Creators of Victoria's London by Stephen Halliday, Breedon Books, 2003
Old London Bridge: the Story of the Longest Inhabited Bridge in Europe by Patricia Pierce, Headline, 2001
'"Rich Earth below the Sand" and the Origins of the Thames Embankments' by Stuart Oliver, *www.reconstruction.ws/023/oliver.htm*
Thames Crossings: Bridges, Tunnels and Ferries by Geoffrey Phillips, David & Charles, 1981
The Thames from Source to Sea by Paul Atterbury and Anthony Haines, The Book People, 1998
Tidal Thames: History of a River and its Fishes by Alwyne Wheeler, Routledge, 1979
Unequal City: London in the Global Arena by Chris Hamnett, Routledge, 2003

CREDITS

Thames map Daniel Morgenstern
Vauxhall Bridge Vanessa Sherry
London Bridge Steve Ibb
Putney Bridge Nikki Kastner
Westminster Bridge Hugh O'Malley
Blackfriars Bridge Christine Marshall
Battersea Bridge Christine Marshall
Waterloo Bridge Alison Locke
Southwark Bridge Paul Wright
Hammersmith Bridge Andrew Buurmann
Hungerford Bridges Naoko Yogo
Chelsea Bridge Paul Wright
Lambeth Bridge Ben Amure
Albert Bridge Heike Lowenstern
Wandsworth Bridge Andrew Buurmann
Tower Bridge Anne Brassier
Millennium Bridge Chris Cooke

INDEX

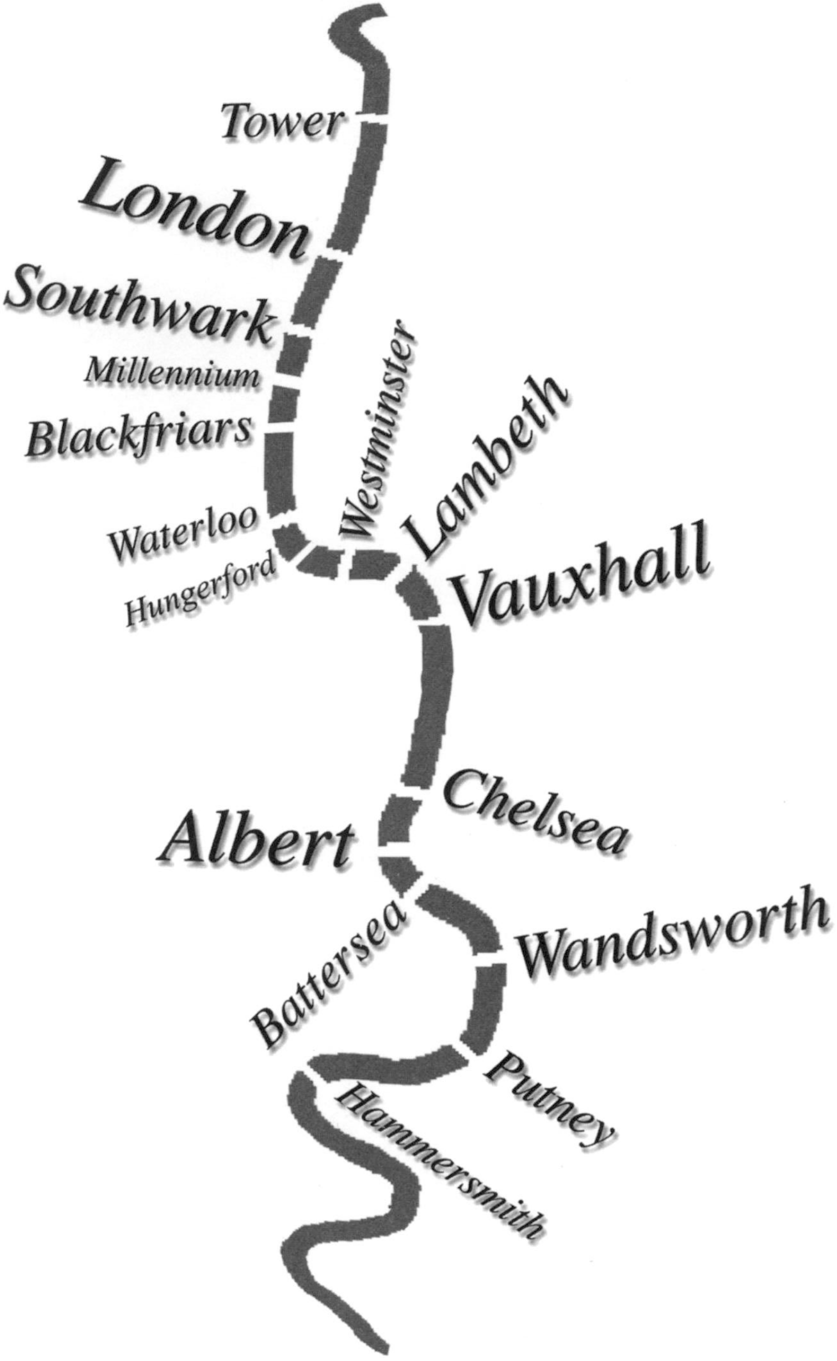

Tower
London
Southwark
Millennium
Blackfriars
Waterloo
Hungerford
Westminster
Lambeth
Vauxhall
Chelsea
Albert
Battersea
Wandsworth
Putney
Hammersmith